V. I. Lebedev

An Introduction to Functional Analysis in Computational Mathematics

Birkhäuser
Boston • Basel • Berlin

V. I. Lebedev
Institute of Numerical Mathematics
Russian Academy of Sciences
Moscow, Russia

Library of Congress Cataloging-in-Publication Data

Lebedev, Viacheslav Ivanovich.
[Funktsional 'nyi analiz i vychislitel'naia matematika. English]
An introduction to functional analysis in computational mathematics / V. I. Lebedev
p. cm.
Includes bibliographical references (p. -) and index.

1. Functional analysis. 2. Mathematical analysis. I. Title.
QA320.L3513 1996 96-33412
515 .7--dc20 CIP

Printed on acid-free paper
 Birkhäuser
Softcover reprint of the hardcover 1st edition 1997

ISBN-13:978-1-4612-8666-0 ISBN-13:978-1-4612-4128-7
DOI:10.1007/978-1-4612-4128-7

Typeset by the author in LaTeX.
Printed and bound by Maple Vail, York, PA.

9 8 7 6 5 4 3 2 1

Table of Contents

Preface to the English Edition

The book contains the methods and bases of functional analysis that are directly adjacent to the problems of numerical mathematics and its applications; they are what one needs for the understanding from a general viewpoint of ideas and methods of computational mathematics and of optimization problems for numerical algorithms. Functional analysis in mathematics is now just the small visible part of the iceberg. Its relief and summit were formed under the influence of this author's personal experience and tastes.

This edition in English contains some additions and changes as compared to the second edition in Russian; discovered errors and misprints had been corrected again here; to the author's distress, they jump incomprehensibly from one edition to another as fleas. The list of literature is far from being complete; just a number of textbooks and monographs published in Russian have been included.

The author is grateful to S. Gerasimova for her help and patience in the complex process of typing the mathematical manuscript while the author corrected, rearranged, supplemented, simplified, generalized, and improved as it seemed to him the book's contents. The author thanks G. Kontarev for the difficult job of translation and V. Klyachin for the excellent figures.

V. Lebedev

Preface to the English Edition

Preface to the First Edition

This book is a yearly course of lectures on functional analysis and numerical mathematics read by the author for a number of years to students of mathematical departments at the Moscow Institute for Physics and Technology. A reader is expected to possess the knowledge in advanced and numerical mathematics, initial value problems for differential equations, elements of the theory of functions of real variables, and methods of linear algebra.

The title of the book is too general; its contents should be regarded as an exposition of the necessary elements of functional analysis and its directions that adjoin directly some problems of numerical mathematics and applications and the investigation of computational mathematics problems. The practice of training students has demonstrated the fruitfulness of such an approach to the investigation of problems in numerical mathematics. At present, a series of good textbooks and monographs is dedicated to functional analysis and applications tightly connected with the theory of partial differential equations and numerical mathematics. A list of these works, although far from exhaustive, is presented in the References. The author used these materials widely. A volume and contents of the course were limited by the training level of the students and by the number of lectures.

The book consists of three chapters. Besides the knowledge on the bases of functional analysis, the book presents elements of the theory of variational equations and generalized solutions including the Vishik–Lax-Milgram theorem, Sobolev spaces, and conceptions on embedding theorems. From the functional standpoint, problems of numerical mathematics are examined such as some extremal problems of approximation theory, theory of numerical integration, some variational methods for the minimization of quadrature functionals, and Galerkin and Ritz methods for finding an approximate solution to operator equations, iteration, and in particular, Chebyshev methods for the solution of operator equations.

For the sake of convenience, the begin and end of a proof is marked, respectively, with the signs □ and ■ ; quantors are also widely used: $\forall$, any and $\exists$, there exists. In order to simplify the proofs, formulations of some statements are presented not for the most general case possible.

V. Lebedev

Preface to the Second Edition

The second edition was subjected to minor revision as compared to the first small edition published by the Department of Numerical Mathematics, USSR Academy of Sciences in 1987. Discovered errors, misprints, and inaccuracies had been corrected and small additions made.

The author is thankful for valuable comments on the book's contents made by N. Bakhvalov, I. Daugavet, G. Kobelkov, and I. Mysovskikh; the author tried to take into account all their comments while preparing the second edition.

V. Lebedev

An Introduction to Functional Analysis in Computational Mathematics

"... *It is impossible to imagine the theory of computations with no Banach spaces, as well as with no computers.*"

S. Sobolev

Chapter 1

Functional Spaces and Problems in the Theory of Approximation

A space is understood in mathematics as a set of any objects (sets of numbers, functions, etc.) with certain relationships established among them, similar to those existing in an elementary three-dimensional space.

Spaces, whose elements are functions or number sequences, are usually called *functional* spaces. Types of functional spaces differ from each other by what property (or properties) of a conventional three-dimensional space is chosen as a relationship (relationships) in a functional space.

§ 1. Metric Spaces

Axioms of a metric space. Opened, closed ball, and a sphere in a metric space. Examples of spaces: $\mathbf{R}^n$, m, l_1, l_2, $C(\overline{\Omega})$, $C^k(\overline{\Omega})$, $\tilde{L}_1(\Omega)$, $\tilde{L}_2(\Omega)$, $\tilde{L}_p(\Omega)$. Convergence, properties of limits, closed and open sets. Fundamental sequences and their properties. Completeness and separability $C[a,b]$, completeness l_2. Theorem on embedded balls. Supplement of spaces. Continuous operators and functionals, abstract functions.

We select for our purposes relationships of the conventional three-dimensional space that are connected with the distances between objects. Let $\mathbf{X}$ be some set, $x, y, z \in \mathbf{X}$.

1. Definition of a metric space

A set $\mathbf{X}$ is called a *metric space* if each couple x and y of its elements has a real number $\rho_{\mathbf{X}}(x,y)$ corresponding to it that satisfies the following conditions (axioms).

(1) $\rho_{\mathbf{X}}(x,y) \geq 0$, $\rho_{\mathbf{X}}(x,y) = 0$ iff $x = y$ (the identity axiom);

(2) $\rho_{\mathbf{X}}(x,y) = \rho_{\mathbf{X}}(y,x)$ (the symmetry axiom);

(3) $\rho_{\mathbf{X}}(x,y) \leq \rho_{\mathbf{X}}(x,z) + \rho_{\mathbf{X}}(z,y)$ (the triangle axiom).

The number $\rho_{\mathbf{X}}(x,y)$ is called a *distance* between the elements x and y (or a *metric* of space $\mathbf{X}$), and conditions (1)–(3) are called the metric axioms. If it is clear what metric space $\mathbf{X}$ is meant, we write $\rho(x,y)$ instead of $\rho_{\mathbf{X}}(x,y)$. Elements of metric space are also called the points.

The so-called triangle inequality follows from the metric axioms

$$|\rho(x,z) - \rho(y,z)| \leq \rho(x,y). \tag{1.1}$$

□ Indeed, it follows from axiom (3), that

$$\rho(x,z) - \rho(y,z) \leq \rho(x,y)$$

and

$$\rho(y,z) - \rho(x,z) \leq \rho(x,y). \blacksquare$$

Any set Y, lying in a metric space $\mathbf{X}$, that has the same distances between elements as $\mathbf{X}$ does, is referred to as a *subspace* of the space $\mathbf{X}$. *Diameter* $d(A)$ of the set $A \subset \mathbf{X}$ is the value

$$d(A) = \sup_{x,y \subset A} \rho(x,y).$$

The set $A \subset \mathbf{X}$ is referred to as *bounded* if its diameter $d(A) < \infty$, which means that an element $x_0 \in \mathbf{X}$ and a constant $c > 0$ can be found such that $\rho(x,x_0) < c$ for $\forall x \in A$.

We call *ball, closed ball*, and *sphere* centered in a point x_0 and of a radius $r > 0$, respectively, the sets of points:

$$\begin{aligned} D_r(x_0) &= \{x \in \mathbf{X} : \rho(x,x_0) < r\} \\ \overline{D}_r(x_0) &= \{x \in \mathbf{X} : \rho(x,x_0) \leq r\} \\ S_r(x_0) &= \{x \in \mathbf{X} : \rho(x,x_0) = r\}. \end{aligned} \tag{1.2}$$

2. Examples of metric spaces

1.1. *Number line.* Let $\mathbf{X} = \mathbf{R}^1$, where $\mathbf{R}^1$ is a set of all real numbers, and $\rho(x, y) = |x - y|$ for $x, y \in \mathbf{R}^1$. We can also take as $\mathbf{X}$ $[a, b]$ or (a, b), where $a \neq b$ are real numbers.

1.2. Let $\mathbf{X}$ be a finite set of elements $x_1, x_2, \ldots, x_n$. Let us assign

$$\rho(x_i, x_j) = \begin{cases} 0, & \text{if} \quad x_i = x_j \\ 1, & \text{if} \quad x_i \neq x_j \,. \end{cases}$$

1.3. *Real n-dimensional space.* Let $\mathbf{X} = \mathbf{R}^n$, $x = (x_1, x_2, \ldots, x_n)$, $y = (y_1, y_2, \ldots, y_n) \in \mathbf{R}^n$. A metric can be introduced in $\mathbf{R}^n$ in a variety of ways, for example:

$$\rho(x, y) = \sum_{i=1}^{n} |x_i - y_i| \tag{1.3}$$

or

$$\rho(x, y) = \sqrt{\sum_{i=1}^{n} (x_i - y_i)^2} \tag{1.4}$$

or

$$\rho(x, y) = \max_i |x_i - y_i|. \tag{1.5}$$

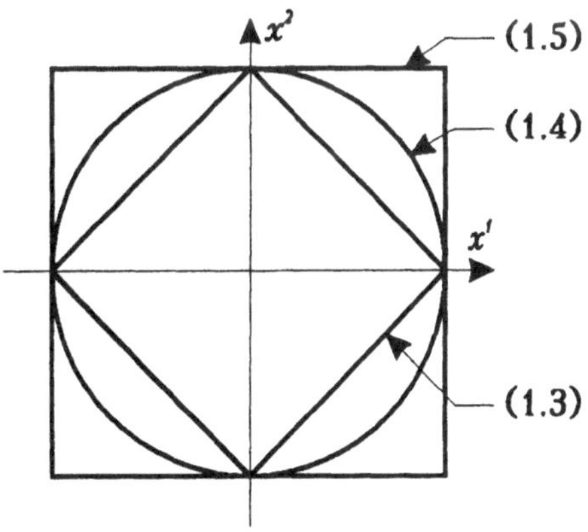

Figure 1

For every metric we obtain its own metric space; a ball and a sphere do not coincide in these three spaces for $r > 0$ (see Figure 1 for $\mathbf{R}^2$).

In Examples **1.4–1.6** $\mathbf{X}$ is a set of infinite number of sequences $x = (x_1, x_2, \ldots, x_n, \ldots)$.

1.4. *Space of number sequences l_1.* Let $x = (x_1, x_2, \ldots, x_n, \ldots)$, $\sum_{i=1}^{\infty} |x_i| < \infty$, and let $y = (y_1, y_2, \ldots, y_n, \ldots)$, $\sum_{i=1}^{\infty} |y_i| < \infty$; assign

$$\rho(x, y) = \sum_{i=1}^{\infty} |x_i - y_i|. \tag{1.6}$$

1.5. *Space of number sequences l_2.* Let $x = (x_1, x_2, \ldots, x_n, \ldots)$ and $\sum\limits_{i=1}^{\infty} x_i^2 < \infty$, and let $y = (y_1, y_2, \ldots, y_n, \ldots)$ and $\sum\limits_{i=1}^{\infty} y_i^2 < \infty$. Assign

$$\rho(x, y) = \sqrt{\sum_{i=1}^{\infty} (x_i - y_i)^2}. \tag{1.7}$$

1.6. *Space of bounded number sequences m (or l_∞).* Let $\mathbf{X}$ be a set of bounded number sequences $x = (x_1, x_2, \ldots, x_n, \ldots)$; it means that for each x there exists $C_x > 0$ such that $|x_i| \leq C_x$ for $\forall i$. Define a distance between elements $x = (x_1, x_2, \ldots, x_n, \ldots)$ and $y = (y_1, y_2, \ldots, y_n, \ldots)$ with the formula

$$\rho(x, y) = \sup_i |x_i - y_i|. \tag{1.8}$$

We call this a space of bounded sequences and denote it by m (or $l_{\nu o}$).

1.7. *Space of continuous Chebyshev functions $C[a, b]$.* Let $\mathbf{X}$ be a set of all continuous functions defined on a segment $[a, b]$; if $f(x)$, $g(x) \in \mathbf{X}$, then assign

$$\rho(f, g) = \max_{x \in [a,b]} |f(x) - g(x)|. \tag{1.9}$$

Similarly, we understand $C(\overline{\Omega})$, where $\Omega \subset \mathbf{R}^n$ is some bounded domain in $\mathbf{R}^n$ and $\overline{\Omega}$ is its closure, as a continuous on $\overline{\Omega}$ functions space with the metric

$$\rho(f, g) = \max_{x \in \overline{\Omega}} |f(x) - g(x)|, \tag{1.10}$$

where $x = (x_1, x_2, \ldots, x_n)$.

1.8. *Space of continuous and kth derivative functions $C^k[a, b]$.* Let $\mathbf{X}$ be a set of all continuous functions defined on an interval $[a, b]$ and having on $[a,b]$ continuous derivatives up to kth order $(k \geq 1)$. If $f(x), g(x) \in \mathbf{X}$, then assign

$$\rho(f,g) = \sum_{i=0}^{k} \max_{x\in[a,b]} |f^{(i)}(x) - g^{(i)}(x)|. \tag{1.11}$$

Similarly, a space $C^k(\overline{\Omega})$ is defined, if a distance is introduced as

$$\rho(f,g) = \sum_{0\leq|\alpha|\leq k} \max_{x\in\overline{\Omega}} |D^\alpha(f-g)|, \tag{1.12}$$

where

$$D^\alpha f(x) = \frac{\partial^{|\alpha|} f(x_1, x_2, \ldots, x_n)}{\partial x_1^{\alpha_1} \partial x_2^{\alpha_2} \ldots \partial x_n^{\alpha_n}}, \qquad D^0 f(x) = f(x)$$

$$D = (D_1, D_2, \ldots, D_n), \qquad D_j = \frac{\partial}{\partial x_j}, \qquad j = 1, 2, \ldots, n$$

and multiindex α is a vector with integer nonnegative components:

$$\alpha = (\alpha_1, \alpha_2, \ldots, \alpha_n), \qquad |\alpha| = \sum \alpha_i, \qquad \alpha_i \geq 0.$$

1.9. *Space of functions that are first order integrable* $\tilde{L}_1(a,b)$. Let $\mathbf{X}$ be a set of all functions continuous on an interval $[a,b]$; if $f(x)$, $g(x) \in \mathbf{X}$, then assign

$$\rho(f,g) = \int_a^b |f(x) - g(x)| dx. \tag{1.13}$$

A space $\tilde{L}_1(\Omega)$ is defined in the same way; here $\Omega \subset \mathbf{R}^n$. For this space

$$\rho(f(x), g(x)) = \int_\Omega |f(x) - g(x)| d\Omega, \qquad d\Omega = dx_1 dx_2 \ldots dx_n. \tag{1.14}$$

1.10. *Space of functions that are second order integrable* $\tilde{L}_2(a,b)$. Let $\mathbf{X}$ be a set of all functions continuous on the interval $[a,b]$; if $f(x)$, $g(x) \in \mathbf{X}$, then assign

$$\rho(f,g) = \left(\int_a^b (f(x) - g(x))^2 dx \right)^{1/2}. \tag{1.15}$$

A space $\tilde{L}_2(\Omega)$ is defined in the same way; here $\Omega \subset \mathbf{R}^n$. For this space

$$\rho(f,g) = \left(\int_{\Omega} (f(x) - g(x))^2 d\Omega \right)^{1/2}. \tag{1.16}$$

1.11. *Space of functions that are p-order integrable* $\tilde{L}_p(a,b)$. Let $p \geq 1$; the spaces $\tilde{L}_1$ and $\tilde{L}_2$ can be generalized as $\tilde{L}_p(a,b)$ and $\tilde{L}_p(\Omega)$. Let us assign on the set of all continuous on the interval $[a,b]$ functions

$$\rho(f,g) = \left(\int_a^b |f(x) - g(x)|^p dx \right)^{1/p}. \tag{1.17}$$

It will be the space $\tilde{L}_p(a,b)$. The space $\tilde{L}_p(\Omega)$ is defined in the same way; here $\Omega \subset \mathbf{R}^n$. For this space

$$\rho(f,g) = \left(\int_{\Omega} |f(x) - g(x)|^p d\Omega \right)^{1/p}. \tag{1.18}$$

Integrals in the spaces $\tilde{L}_1$, $\tilde{L}_2$, and $\tilde{L}_p$ are understood in the Riemann sense.

It is possible to verify the validity of the metric axioms for all spaces introduced here. The triangle axiom for the spaces $\tilde{L}_p(a,b)$ and $\tilde{L}_p(\Omega)$ appears as the Minkowskii inequality for the integrals

$$\left(\int_{\Omega} |f(x) + g(x)|^p d\Omega \right)^{1/p}$$

$$\leq \left(\int_{\Omega} |f(x)|^p d\Omega \right)^{1/p} + \left(\int_{\Omega} |g(x)|^p d\Omega \right)^{1/p}. \tag{1.19}$$

1.12. Let $w(x) \geq c > 0$ be a continuous function on $[a,b]$ [or on $\bar{\Omega}$, if $x = (x_1, \ldots, x_n)$], that is called a weight. Let us introduce a function $w(x)$ into formulas for distances (1.9)–(1.18) as a multiplier, to obtain new spaces $C([a,b],w)$, if instead of (1.9) we take

$$\rho(f,g) = \max_{x \in [a,b]} |(f(x) - g(x))w(x)| \tag{1.9'}$$

or $\tilde{L}_p(\bar{\Omega}, w)$, if instead of (1.18) we take

$$\rho(f,g) = \left(\int_{\Omega} |f(x) - g(x)|^p w(x)\, d\Omega \right)^{1/p} \tag{1.18'}$$

and so forth.

Since the statements concerning spaces with a weight or without ($w(x) \equiv 1$) stay unchanged, we deal, as a rule, with the case $w(x) \equiv 1$.

Spaces (1.3)–(1.7) and (1.9)–(1.11) exemplify that a metric may be introduced on the same set $\mathbf{X}$ in different ways, which results, generally speaking, in different metric spaces. To be more accurate, *we call the metric space a couple* $(\mathbf{X}, \rho)$, *that consists of the space* $\mathbf{X}$ *and the metric* ρ.

3. Convergence in a metric space, limit of a sequence

Definition. A point x of metric space $\mathbf{X}$ is called a *limit of infinite sequence* of points $x_n \in \mathbf{X}$ (we write $x_n \to x$ or $x = \lim x_n$), if $\rho(x_n, x) \to 0$ for $n \to \infty$. Convergence of a sequence $x_1, x_2, \ldots, x_n, \ldots$ so defined is called the *distance (metric) convergence of the space* $\mathbf{X}$.

It is possible to make a number of simple statements concerning convergent sequences.

Lemma 1. *A sequence of points* $\{x_n\}$ *of a metric space converges to only one limit.*

□ Let $x_n \to x$, $x_n \to y$; by the triangle axiom $\rho(x,y) \le \rho(x_n, x) + \rho(x_n, y) \to 0$ for $n \to \infty$; therefore, $\rho(x,y) = 0$; that is, $x = y$. ■

Lemma 2. *A distance* $\rho(x,y)$ *is a continuous function of its arguments; that is, if* $x_n \to x$, $y_n \to y$, *then* $\rho(x_n, y_n) \to \rho(x,y)$ *for* $n \to \infty$.

□ Each time, using twice the triangle axiom inequality we obtain for $n \to \infty$

$$|\rho(x_n, y_n) - \rho(x,y)| \le \rho(x_n, x) + \rho(y_n, y) \to 0. \ ■$$

4. Closed and open sets, closure of sets, neighborhood, distance to a set

Consider a set $A \subset \mathbf{X}$. A point $x \in A$ is called an *internal point of the set* A if $\exists \varepsilon > 0$ such that $D_\varepsilon(x) \subset A$; the point $x \in \mathbf{X}$ is called a *limit point of the set* A if $\forall \varepsilon > 0$ $D_\varepsilon(x) \cap (A \backslash x) \neq \emptyset$; that is, there exists a sequence $u_n \in A \setminus x$ such that $u_n \to x$. A set obtained by joining all its limit points to A is called a *closure of the set* A and denoted by $\bar{A}$. A set A is referred to as *closed* if $\bar{A} = A$. A set A is termed *open* if all its points are internal.

A set $A \subset \mathbf{X}$ is termed *dense* in the set $N \subset \mathbf{X}$ if $\bar{A} \supseteq N$. A set A is referred to as *everywhere dense* in the space $\mathbf{X}$ if $\bar{A} = \mathbf{X}$.

A *neighborhood* of a point x is any open set containing the point x, for example, any ball $D_\varepsilon(x)$.

A *distance* from a point x to a set A is the number

$$\rho(x, A) = \inf_{y \in A} \rho(x, y). \tag{1.20}$$

The set

$$O_\varepsilon(A) = \{x \in \mathbf{X} : \rho(x, A) < \varepsilon\} \tag{1.21}$$

is called an *ε-neighborhood* of the set A.

Lemma 3. *If $x \in A$, then $\rho(x, A) = 0$. If $x \notin A$ and A is closed, then $\rho(x, A) > 0$.*

□ If $x \in A$, then for $y = x$ we have $\rho(x, A) = 0$. Now let $x \notin A$, and let A be closed. Let us take that $\rho(x, A) = 0$.

By the definition of a lower bound, $u_n \in A$ will be found for $\forall\, n$ such that $\rho(x, u_n) < 1/n$. Therefore, $u_n \to x$ for $n \to \infty$ and $x \in A$ due to the closure of A, but by condition $x \notin A$. The obtained contradiction proves the lemma. ■

Remark 1. The concepts of open set and neighborhood were defined with the help of a metric assigned on $\mathbf{X}$. It is possible, using the axioms, to directly define in $\mathbf{X}$ some system of sets, having termed them open (introducing them shelf by shelf), and call, as before, a neighborhood of a point $x \in \mathbf{X}$ any open set containing the point x. This way leads to topological spaces, whose special case is metric spaces.

5. Convergent in itself or fundamental sequences

A sequence $\{x_n\}$ of elements of metric space $\mathbf{X}$ is called a *convergent in itself (fundamental)* sequence, if for $\forall\varepsilon > 0$ $\exists n_0(\varepsilon)$ such that $\rho(x_n, x_m) < \varepsilon$ for $n, m \geq n_0(\varepsilon)$.

Theorem 1. *If a sequence $\{x_n\}$ converges to a limit x_0, then it is fundamental.*

□ Let $x_n \to x_0$; then $\rho(x_n, x_m) \leq \rho(x_n, x_0) + \rho(x_0, x_m) \to 0$ for $n, m \to \infty$. ■

The inverse statement for an arbitrary metric space is, generally speaking, incorrect (for example, if $\mathbf{X}$ is a set of all rational numbers).

Theorem 2. *Any fundamental sequence $\{x_n\}$ is bounded.*

□ Let us assign $\varepsilon > 0$ and select N so that $\rho(x_n, x_m) < \varepsilon$ for $n, m \geq N$. Then $\rho(x_n, x_N) < \varepsilon$ for $n \geq N$. Let $r = \max\{\varepsilon, \rho(x_i, x_N), i = 1, 2, \ldots, N-1\}$; then for $\forall\, n,\ \ x_n \in \overline{D}_r(x_N)$. ■

6. Complete spaces

Let us present the main definition which in the theory of metric spaces. A metric space $\mathbf{X}$ is referred to as *complete* if any fundamental sequence in this space has a limit that belongs to this space. Most of the preceding examples of spaces are complete.

Spaces $\mathbf{R}^n$ with introduced metrics are complete by virtue of the Cauchy criterion on the existence of a limit for the points of these spaces.

Lemma 4. *Spaces $C[a, b]$, $C(\overline{\Omega})$ are complete.*

□ Let a sequence $f_n(x) \in C[a, b]$ be such that $\rho(f_n, f_m) \to 0$ for $n, m \to \infty$; that is, the Cauchy condition on uniform convergence on $[a, b]$ is satisfied for the sequence $f_n(x)$. Therefore, $\exists\ f_0(x) \in C[a, b]$, to which the sequence $f_n(x)$ converges uniformly on $[a, b]$. Completeness of $C(\overline{\Omega})$ is proven in the same way. ■

Lemma 5. *The space l_2 is complete.*

□ Let a sequence of elements $x^m = (x_1^m, x_2^m, \ldots, x_n^m, \ldots) \in l_2$ be such that $\rho(x^m, x^p) \to 0$ for $m, p \to \infty$. However,

$$\rho^2(x^m, x^p) = \sum_{i=1}^{\infty} (x_i^m - x_i^p)^2$$

therefore, $x_i^m - x_i^p \to 0$ for each $i = 1, 2, \ldots$ It follows from the Boltsano–Cauchy criterion for real numbers that $\lim_{m\to\infty} x_i^m = x_i$ for each i. Let us show that the vector $x = (x_1, x_2, \ldots, x_n, \ldots) \in l_2$. By the assumption for $\forall\ \varepsilon > 0\ \exists\ M > 0$ such that for $m, p \geq M$

$$\sum_{i=1}^{\infty} (x_i^m - x_i^p)^2 < \frac{1}{2}\varepsilon^2$$

the more so for $\forall\ n \geq 1$

$$\sum_{i=1}^{n} (x_i^m - x_i^p)^2 < \frac{1}{2}\varepsilon^2.$$

Fix m and pass to a limit for $p \to \infty$ to obtain

$$\sum_{i=1}^{n} (x_i^m - x_i)^2 \leq \frac{1}{2}\varepsilon^2.$$

Now also direct n to infinity to obtain

$$\sum_{i=1}^{\infty} (x_i^m - x_i)^2 \leq \frac{1}{2}\varepsilon^2 < \varepsilon^2;$$

that is, $x^m - x \in l_2$ and $\rho(x^m, x) < \varepsilon$ for $m > M$, but then $x = x^m + (x - x^m) \in l_2$ too; in addition $x^m \to x$. ■

It is possible to prove that the spaces $\mathbf{R}^n$, $C^k[a,b]$, $C^k(\overline{\Omega})$, m, l_1 are complete, and the spaces $\tilde{L}(a,b)$, $\tilde{L}_1(\Omega)$, $\tilde{L}_2(a,b)$, $\tilde{L}_2(\Omega)$, $\tilde{L}_p(a,b)$, $\tilde{L}_p(\Omega)$ are incomplete. The incompleteness of the latter is proven easily by the construction of fundamental sequences of continuous functions which converge in the mean (respectively, in square mean and in pth order mean) to the discontinuous function.

Theorem 3 (on embedded balls). Let a complete metric space **X** *contain a sequence of closed balls $\overline{D}_{\varepsilon_n}(a_n)$, $n = 1, 2, \ldots$ embedded in each other [i.e., $\overline{D}_{\varepsilon_i}(a_i) \subset \overline{D}_{\varepsilon_k}(a_k)$ for $i > k$], whose radii tend to zero. Then there exists one and only one point $\bar{a} \in \mathbf{X}$ that belongs to these balls.*

□ By definition, $\varepsilon_n \to 0$ for $n \to \infty$. But, since for $m > n$ $a_m \in \overline{D}_{\varepsilon_m}(a_m) \subset \overline{D}_{\varepsilon_n}(a_n)$, hence $\rho(a_n, a_m) \leq \varepsilon_n$; that is, the sequence a_n is fundamental. There exists $a = \lim a_n \in \mathbf{X}$ by virtue of completeness of **X**. Since $\{a_{n+p}\} \in \overline{D}_{\varepsilon_n}(a_n)$ for $\forall n, p = 1, 2, \ldots, a_{n+p} \to a$ for $p \to \infty$ and the ball $\overline{D}_{\varepsilon_n}(a_n)$ is closed, then $a \in \overline{D}_{\varepsilon_n}(a_n)$ for $\forall\ n = 1, 2, \ldots$. If there exists another point b ($a \neq b$) that belongs to all balls, then $\rho(a, b) = \delta > 0$ and, on the other hand, we would have

$$\delta = \rho(a, b) \leq \rho(a, a_n) + \rho(a_n, b) \leq 2\varepsilon_n \to 0$$

which is impossible. ■

7. Isometry and separability

Two metric spaces $\mathbf{X}, \mathbf{Y}$ are termed *isometric* if one-to-one correspondence can be established between two of their elements $x, x' \in \mathbf{X}$, $y, y' \in \mathbf{Y}$, so that

$$\rho_{\mathbf{X}}(x, x') = \rho_{\mathbf{Y}}(y, y')$$

It is possible, from the standpoint of convergence, completeness, and so on, to consider two isometric spaces as identical.

A metric space is termed *separable* if there exists in a it countable or finite everywhere dense set.

All the spaces considered in the preceding examples are separable except for the space m. For instance, the set $C[a, b]$ is separable, since it contains a set $\mathcal{P}$ of all algebraic polynomials with rational coefficients which is a countable everywhere dense set.

In fact, by the Weierstrass theorem (see Section 9 of this chapter) every continuous on $[a, b]$ function $f(x)$ can be approximated with arbitrary accuracy by a uniformly convergent sequence of polynomials $P_n(x)$; that is, $P_n(x) \to f(x)$ in the metric $C[a, b]$. Having substituted all coefficients of each polynomial $P_n(x)$ with close enough rational ones, we will find the polynomials $Q_n(x) \in \mathcal{P}$ such that

$$\max_{x \in [a,b]} |P_n(x) - Q_n(x)| < \frac{1}{n};$$

that is, $\rho(P_n, Q_n) < (1/n)$. Therefore, $Q_n(x) \to f(x)$ which means that the set $\mathcal{P}$ is everywhere dense in $C[a,b]$.

Since convergence in $C[a,b]$ implies convergence in $\tilde{L}_p(a,b)$, then the set $\mathcal{P}$ is everywhere dense in $\tilde{L}_p(a,b)$ and, therefore, these spaces are separable. The following lemma is valid.

Lemma 6. *If everywhere dense in* $\mathbf{X}$ *subset* M *of metric space* $\mathbf{X}$ *is a separable space, then* $\mathbf{X}$ *is separable too.*

□ Let A be the countable everywhere dense subset M. Take $x \in \mathbf{X}$ and assign $\varepsilon > 0$. Since M is everywhere dense in $\mathbf{X}$, then $\exists\ x' \in M$ such that $\rho(x, x') < (\varepsilon/2)$, and since A is everywhere dense in M, then $\exists\ x'' \in A$ such that $\rho(x', x'') < (\varepsilon/2)$; therefore, $\rho(x, x'') < \varepsilon$ and, since ε is arbitrary, this means that A is everywhere dense in $\mathbf{X}$. ■

8. Complement of metric spaces

Let us analyze the complement process of arbitrary incomplete metric space, which is similar to the supplement process for rational numbers. Let an incomplete metric space $\mathbf{X}$ be given. It means that there exist fundamental sequences, composed of the elements of this space, having no limit in $\mathbf{X}$. The following statement is valid: *any metric space can be complemented; or, more correctly: there exists another complete metric space* $\mathbf{X}'$ *such that there exists in it everywhere complete in* $\mathbf{X}'$ *subspace* $\mathbf{X}_0$ *which is isometric to the space* $\mathbf{X}$. The space $\mathbf{X}'$ is called a *complement* of the space $\mathbf{X}$.

We do not prove this statement, referring the reader to textbooks on functional analysis. It is the same as the proof of the theorem on the supplement of sets of rational numbers to the set of real numbers with classes of equivalent fundamental sequences.

Example 1.13. Let $\mathbf{X}$ be a space of polynomials defined on $[a,b]$ with the metrics $\rho(p,q) = \max\limits_{x\in[a,b]} |p(x) - q(x)|$. It is obvious, that this space is incomplete. Since $\mathbf{X}$ lies everywhere densely in complete space $C[a,b]$ (by the Weierstrass theorem), then the supplement of $\mathbf{X}$ is $C[a,b]$ (more correctly, a space isometric to $C[a,b]$).

Let us denote the supplements of spaces $\tilde{L}_1(a,b)$, $\tilde{L}_1(\Omega)$, $\tilde{L}_2(a,b)$, $\tilde{L}_2(\Omega)$, $\tilde{L}_p(a,b)$, $\tilde{L}_p(\Omega)$, respectively, by $L_1(a,b)$, $L_1(\Omega)$, $L_2(a,b)$, $L_2(\Omega)$, $L_p(a,b)$, $L_p(\Omega)$. These spaces are called *Lebesgue spaces.*

Elements of these spaces may be considered as some ideal functions that can always be approximated in a corresponding metric with the help of fundamental sequences composed of continuous functions. It is possible to identify some elements of $L_p(\Omega)$ as some concrete, generally speaking, discontinuous functions. By definition, the limit

$$\int_\Omega f(x)\,d\Omega = \lim_{n\to\infty} \int_\Omega f_n(x)\,d\Omega \tag{1.22}$$

will be called the *Lebesgue integral* of a function $f(x) \in L_p(\Omega)$. [The Lebesgue integral is on the left-hand side of (1.22), and the Riemann integral is on the right-hand side.] Here $f_n(x) \in the\tilde{L}_p(\Omega)$ is any fundamental in $L_p(\Omega)$ sequence, converging to $f(x)$.

Let us show that the sequence of definite integrals in the right-hand side of (1.22) is fundamental.

□ We have for $f_n(x), f_m(x) \in \tilde{L}_p(\Omega)$ that

$$\rho_{n,m} = \left| \int_a^b f_n(x)\,dx - \int_a^b f_m(x)\,dx \right| \le \int_a^b |\, f_n(x) - f_m(x)|\,d\Omega. \tag{1.23}$$

If $p > 1$, then in order to estimate ρ_{nm} we use the *Hölder inequality*: for $\forall\, f(x), g(x) \in C[\Omega]$

$$\left| \int_a^b f(x)g(x)\,dx \right| \le \left[\int_a^b |f(x)|^p\,dx \right]^{1/p} \left[\int_a^b |g(x)|^q\,dx \right]^{1/q}, \tag{1.24}$$

where $q^{-1} = 1 - p^{-1}$. Then, assigning $f(x) = f_n(x) - f_m(x)$, $g(x) = 1$ we obtain from (1.23)

$$\rho_{n,m} \le |\int_\Omega d\Omega|^{1/q} \left[\int_\Omega |f_n(x) - f_m(x)|^p\,d\Omega \right]^{1/p}. \tag{1.25}$$

Therefore, for $\forall\, \varepsilon > 0\ \exists\, n_0$ such that for $n, m > n_0\ \ \rho_{nm} < \varepsilon$, it follows from (1.23) for $p = 1$ and from (1.25) for $p > 1$. ■

Remark 2. A reader not familiar with the theory of measure and measurable functions may, at first, accept this nonstrict definition of the Lebesgue integral. While formulating it, we had "combed" the set of integrable functions, neglecting: (a) classes of functions from $L_p(\Omega)$ that are equivalent to an assigned function and differ from it just on a set of zero measure; (b) investigation of the convergence of sequences composed of measurable functions as the convergence almost everywhere. Presentation of Lebesgue integral theory would take a dozen pages.

We hereinafter regard a metric space as a complete metric space if it is not specially stipulated.

9. Functional dependence, continuous operators in a metric space

Let $\mathbf{X}, \mathbf{Y}$ be two metric spaces, $D \subset \mathbf{X}$. If every element of $x \in D$ is correlated according to certain rules to the unique and fully definite element $y \in Y$, then we say that an *operator* P [notation $y = Px$ or $y = P(x)$ is possible too] with a domain of values that belongs to the space $\mathbf{Y}$ is assigned on the set D. Sometimes we also say that a *mapping* of the set $D \subset \mathbf{X}$ into the set $\mathbf{Y}$ is defined and write $f : D \to \mathbf{Y}$. If values of an operator are real or complex numbers, then the operator is called a *functional.*

If $\mathbf{X} = \mathbf{R}^1$ is the number line, then the mapping $P(x)$ is called an *abstract function.*

An operator $P(x)$ is referred to as *continuous* in the point $x_0 \in D$, if for $\forall \varepsilon > 0$ $\exists \delta > 0$ such that $\rho_{\mathbf{Y}}(P(x), P(x_0)) < \varepsilon$ for $\forall\, x \in D$ from the ball $\rho_{\mathbf{X}}(x, x_0) < \delta$. It follows from the definition of continuity for $P(x)$ that if $x_n \to x_0$ $(x_n, x_0 \in D)$, then $P(x_n) \to P(x_0)$. It is easily shown, that also on the contrary, if indicated limit takes place for any sequence $x_n \to x_0$, then $P(x)$ is continuous in the point x_0. An operator $P(x)$ is termed continuous on D, if it is continuous in any point $x \in D$. We say that $P(x)$ satisfies on D the *Lipschitz condition* with a constant $L > 0$, if for $\forall\, x_1, x_2 \in D$ the condition

$$\rho_{\mathbf{Y}}(P(x_1), P(x_2)) \leq L\rho_{\mathbf{X}}(x_1, x_2) \tag{1.26}$$

is fulfilled. It is obvious, that if condition (1.23) is satisfied, then the operator P is continuous on D. Additional properties of such operators for $L < 1$ are investigated in Section 4 of this chapter.

§ 2. Compact Sets in Metric Spaces

Compact, bicompact, locally compact spaces. Boundedness and compactness. ε-net. Hausdorff theorem. ε-Entropy. Bicompact sets and problems of calculus of variations.

The following Boltsano–Weierstrass theorem is one of the most important theorems of mathematical analysis: *it is possible to separate in any bounded infinite sequence a subsequence converging to the finite limit.* Let us separate a class of sets lying in a metric space, for whom the conclusion of this theorem is valid. Let $\mathbf{X}$ be a metric space, and let a set $M \subset \mathbf{X}$.

1. Bicompact, compact, locally compact sets and their properties

A set $M \subset \mathbf{X}$ is termed *bicompact* if it is possible to select in each sequence $\{x_n\} \subset M$ a subsequence whose limit belongs to M.

A set $M \subset \mathbf{X}$ is termed *compact* if it is possible to separate in each sequence $\{x_n\} \subset M$ a fundamental subsequence.

An unbounded set M is referred to as *locally compact* (*bicompact*) if an intersection of M with any closed ball is compact (bicompact).

The notion of a compact set is weaker, by virtue of the preceding definitions, than the notion of a bicompact set, since the following corollaries and lemmas are valid.

Corollary 1. *Any bicompact set M is compact.*

Corollary 2. *Any bicompact set M is closed.*

Therefore, the following lemma is valid.

Lemma 1. *Compact set M is bicompact iff it is closed.*

Lemma 2. *Any compact set M is bounded.*

□ We perform the proof by contradiction. Let M be unbounded. Take any of its points and denote it as x_1. Since M is unbounded, the set $M_1 = M \backslash D_1(x_1)$ is not empty.

Let x_2 be any point of M_1, then $\rho(x_1, x_2) \geq 1$. Let $M_2 = M_1 \backslash D_1(x_2)$ which is not an empty set; let $x_3 \in M_2$; then $\rho(x_1, x_3) \geq 1$, $\rho(x_2, x_3) \geq 1$. Continuing this process infinitely, we obtain a sequence of points $\{x_n\} \in M$ such that $\rho(x_i, x_n) \geq 1$ for $i \neq n$. Therefore, no partial sequence separated from $\{x_n\}$ can be fundamental; more than that, it will not converge to a limit, which means that M is not compact. The contradiction proves the lemma. ∎

The inverse statement is, generally speaking, incorrect, for example in l_2. In fact, all coordinate basis vectors in l_2 $e_m = (0, 0, \ldots, 1, 0, \ldots, 0, \ldots)$ belong to bounded set $\overline{D}_1(e_0)$ where $e_0 = (0, 0, \ldots, 0, \ldots)$. On the other hand, $\rho(e_m, e_p) = \sqrt{2}$ for $p \neq m$, therefore, it is impossible to separate a converging sequence from the sequence e_n. However, the finite-dimensional variant of the Boltsano–Weierstrass *theorem* is proven in the courses of mathematical analysis: *boundedness is the necessary and sufficient condition for a set $M \subset \mathbf{R}^n$ to be compact* (see Section 13 of Chapter 2).

Let $\varepsilon > 0$. A set $M_\varepsilon \subset \mathbf{X}$ is called the ε-*net* of a set M, if a point $\tilde{x} \in M_\varepsilon$ can be found for $\forall\, x \in M$ such that $\rho(x, \tilde{x}) < \varepsilon$. It is evident that

$$M \subset \bigcup_{\tilde{x} \in M_\varepsilon} D_\varepsilon(\tilde{x}); \tag{2.1}$$

that is, the aggregate of balls $D_\varepsilon(\tilde{x})$ covers M. We say that the ε-*net is finite* if M_ε is a finite set of elements.

Theorem 1 (by Hausdorff). *A set M in metric space* $\mathbf{X}$ *is compact iff $\forall\, \varepsilon > 0$ in* $\mathbf{X}$ *there exists a finite ε-net.*

Necessity. Let M be compact; take $\forall \varepsilon > 0$. Let us prove the existence of a finite ε-net reasoning as follows. Take $x_1 \in M$; if $\rho(x, x_1) < \varepsilon$ for $\forall\, x \in M$, then $M_\varepsilon = \{x_1\}$; otherwise $\exists\, x_2 \in M$ such that $\rho(x_1, x_2) \geq \varepsilon$. If $\rho(x, x_1) < \varepsilon$ or $\rho(x, x_2) < \varepsilon$ for $\forall\, x \in M$, then $M_\varepsilon = \{x_1, x_2\}$. Proceeding in this way we see that either the process of construction of x_i will stop at the mth step and we obtain a finite ε-net: $M_\varepsilon = \{x_1, x_2, \ldots, x_m\}$, or the process will continue infinitely. We obtain in the last case $\{x_n\} \subset M$ such that $\rho(x_n, x_m) \geq \varepsilon$ for $\forall\, n, m$ $(n \neq m)$; this sequence and any of its subsequences are not fundamental which contradicts the compactness of M. Hence, the process cannot be infinite. It means that a finite ε-net exists and it can consist of elements of the set M.

Sufficiency. Let there exists for $\forall\, \varepsilon > 0$ a finite ε-net for a set M. Take a sequence $\varepsilon_n \to 0$ $(\varepsilon_n > 0)$ and let

$$M_{\varepsilon_n} = \{x_{n1}, x_{n2}, \ldots, x_{nk_n}\}$$

(the set M_{ε_n} consists of k_n elements). Take any sequence $\{x_n\} \subset M$ nd separate from it a fundamental subsequence with a special method. Recall to this end, that for $\forall\ \varepsilon_n > 0$, $M \subset \bigcup_{i=1}^{k_n} D_{\varepsilon_n}(x_{ni})$.

Since $\{x_n\}$ consists of an infinite number of elements, then there exists $D_{\varepsilon_1}(x_{1i})$ such that it contains an infinite number of members of $\{x_n\}$. Remove from $\{x_n\}$ the members that do not belong to $D_{\varepsilon_1}(x_{1i})$. Denote the sequence that is left (preserving the order of sequence) through $\{x_n^1\}$. Repeat the process to obtain that there exists $D_{\varepsilon_2}(x_{2i})$ such that it contains infinite number of elements from $\{x_n^1\}$. Remove from $\{x_n^1\}$, according to the same rule, the members that do not belong to $D_{\varepsilon_2}(x_{2i})$ and denote through $\{x_n^2\}$ the subsequence that is left and so on. Now the sequence $\{x_n^n\}$ is fundamental since $\rho(x_m^m, x_p^p) < \varepsilon_n$ for $m, p \geq n$. ■

Corollary 3. *If for $\forall\ \varepsilon > 0$ for a set M there exists a compact ε-net in $\mathbf{X}$, then M is compact.*

□ Take $\varepsilon > 0$ and let M_ε be a compact net for the ε-net for M. There exists, by the Hausdorff theorem, a finite ε-net M'_ε for M_ε. It is 2ε-net for M. ■

Corollary 4. *Compact space $\mathbf{X}$ is separable.*

□ Take a sequence $\varepsilon_n \to 0$ $(\varepsilon_n > 0)$ and construct for each ε_n a finite ε-net $M_{\varepsilon_n} = \{x_i^n\}$, $i = 1, 2, \ldots, k_n$. Then a set $M_1 = \bigcup_{n=1}^{\infty} M_{\varepsilon_n}$ is a countable everywhere dense in $\mathbf{X}$ set. ■

Let $\{M_{\alpha_\varepsilon}\}$ be a set of all ε-nets for a set M, N_{α_ε} be the number of elements of ε-net M_{α_ε} for assigned ε and let

$$N_\varepsilon^0 = \inf_{M_{\alpha_\varepsilon}} N_{\alpha_\varepsilon}. \tag{2.2}$$

Then the value

$$H_\varepsilon(M) = \log_2 N_\varepsilon^0 \tag{2.3}$$

is called *ε-entropy of the set M*. The ε-entropy serves as a measure of "information amount" when certain element $x \in M$ is indicated with an accuracy ε, since in order to separate one definite element among N_ε^0 elements of the minimal ε-net or, what is the same, among the numbers $1, \ldots, N_\varepsilon^0$ it is sufficient to assign $[\log_2 N_\varepsilon^0]$ binary signs.

2. Bicompact sets and problems of calculus of variations

There are theorems of mathematical analysis stating that any continuous on an interval function $f(x)$ is bounded and reaches its maximal and minimal values there. The following theorem is valid for metric spaces.

Theorem 2. Let $f(x)$ be a real continuous functional defined on bicompact set $M \subset \mathbf{X}$; then $f(x)$ is bounded on M.

□ Let us prove, for the sake of definiteness, that $f(x)$ is bounded above. Assume the opposite; then $x_n \in M$ will be found such that $f(x_n) > n$. Separate from $\{x_n\}$ the converging sequence $\{x_{n_k}\}$. Let $x_{n_k} \to x_0$ for $k \to \infty$; $x_0 \in M$ since M is bicompact. Continuity of the functional $f(x)$ implies that $f(x_{n_k}) \to f(x_0)$ for $k \to \infty$ where $f(x_0)$ is a certain number. Therefore, $\{f(x_{n_k})\}$ is bounded. On the other hand, $f(x_{n_k}) > n_k$; that is, $f(x_{n_k}) \to \infty$ for $k \to \infty$. It is a contradiction, hence the assumption that $f(x)$ is not bounded above is incorrect, which means that $f(x)$ is bounded above. The boundedness below for $f(x)$ is proven in the same way. ■

Theorem 3. Let $f(x)$ be a real continuous functional defined on bicompact set $M \subset \mathbf{X}$. Then there exist $x_0 \in M$, $x^0 \in M$ such that

$$f(x_0) = \inf_{x \in M} f(x), \qquad f(x^0) = \sup_{x \in M} f(x);$$

that is, $f(x)$ reaches on M its minimal and maximal values.

□ Denote $\sup\limits_{x \in M} f(x) = m$; we have by definition of sup: for $\forall\, n$ $\exists\, x_n \in M$ such that $m - 1/n < f(x_n) \le m$; that is, $f(x_n) \to m$. There exists, by virtue of the bicompactness of M, a subsequence $\{x_{n_k}\}$ converging to some $x^0 \in M$. By continuity $f(x_{n_k}) \to f(x^0)$ for $k \to \infty$. The uniqueness of a limit implies that $m = f(x^0)$. We prove in a similar way that $f(x)$ reaches its least value on M. ■

We obtain new facts on compact sets while studying totally continuous or compact operators (see Section 13 of Chapter 2).

"Study the classics of mathematics and solve difficult problems."

P. Chebyshev

§ 3. Statement of the Main Extremal Problems in the Theory of Approximation. Main Characteristics of the Best Approximations

Problem I on the best approximation of individual element $x \in \mathbf{X}$ by fixed approximating set A. Problem II on the best approximation of assigned set $C \in \mathbf{X}$ by fixed approximating set. Problem III on the best approximation of assigned set C by class $\mathcal{A}$ of approximating sets A. Kolmogorov N-width and ε-entropy of a set C, Chebyshev center. Problem IV on the approximation of assigned set with the help of the fixed approximation method. Problem on the best approximation method. Examples.

Chebyshev's works are the origin of the theory of approximation as an independent part of mathematics, although separate problems were studied by Euler, Gauss, Legendre, Weierstrass, and other mathematicians.

Let $\mathbf{X}$ be a metric space with a distance $\rho(x, y)$, $x, y \in \in \mathbf{X}$ and let $A \subset \mathbf{X}$.

Problem I *on the best approximation of individual element* $x \in \mathbf{X}$ *by fixed approximating set* A.

The task is to find a value (Figure 2)

$$e(x, A, \mathbf{X}) = \inf_{\xi \in A} \rho(\xi, x) \tag{3.1}$$

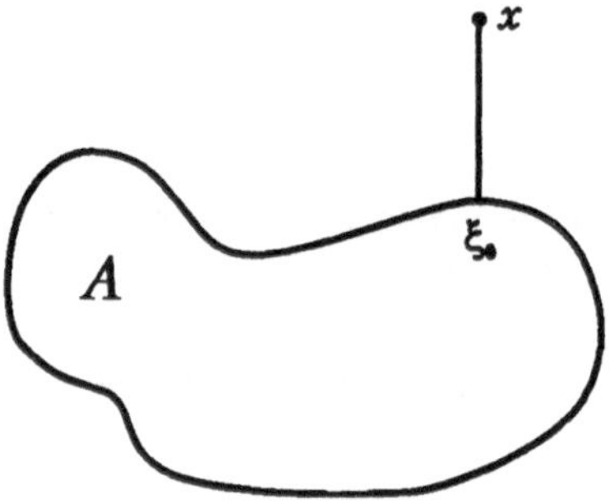

Figure 2

which is equal to the distance between x and the set A and is called *the best approximation*; this value is also denoted as $\rho(x, A)$. An element $\xi_0 \in A$, if it exists, and for which $e(x, A, \mathbf{X}) = \rho(x, \xi_0)$, is called an *element of the best approximation for* x or *extremal element*:

$$\xi_0 = \arg \inf_{\xi \in A} \rho(\xi, x) \tag{3.2}$$

which means that ξ_0 is an argument of $\rho(\xi, x)$ for which the inf is reached. Problem I belongs to problems of mathematical programming.

Example 3.1. $\mathbf{X} = C[a, b]$, $x(t) \in C[a, b]$, $A = A_n$ is a set of polynomials $P_n(t)$ of a power not higher than n:

$$A_n = \{P_n(t) : P_n(t) = \sum_{k=0}^{n} a_k t^k\}. \tag{3.3}$$

It is possible to prove that the extremal polynomial $P_n^0(t)$ always exists and is unique; it is called the *polynomial of the best approximation* for the function $x(t)$. Generally speaking, it is difficult to find such a polynomial (see Sections 9 and 10 of this chapter).

Example 3.2. $\mathbf{X} = L_2(-\pi, \pi)$, $x(t) \in L_2(-\pi, \pi)$, $A = A_n$ is a set of trigonometrical polynomials $Q_n(t)$ of a power not higher than n:

$$A_n = \{Q_n(t) : Q_n(t) = \sum_{k=0}^{n} (a_k \cos kt + b_k \sin kt)\}. \tag{3.4}$$

Here, the extremal polynomial $S_n(t)$ or *the least deviating in average polynomial* from the function $x(t)$, is the polynomial where coefficients a_k, b_k are Fourier coefficients of the function $x(t)$

$$a_0 = \frac{1}{2\pi} \int_{-\pi}^{\pi} x(t)\, dt, \qquad a_k = \frac{1}{\pi} \int_{-\pi}^{\pi} x(t) \cos kt\, dt,$$

$$b_k = \frac{1}{\pi} \int_{-\pi}^{\pi} x(t) \sin kt\, dt, \qquad k = 1, 2, \ldots, n. \tag{3.5}$$

The extremal polynomial of Example 3.2 is assigned with explicit formulas (see Section 8 of this chapter).

Example 3.3. Let $x \in \mathbf{X}$ and let $A = S_\varepsilon(x)$ be a sphere of a radius ε centered in a point x. Then any element $\xi \in S_\varepsilon(x)$ will be the element of the best approximation for x.

Problem II *on the best approximation of assigned set* $C \in \mathbf{X}$ *by fixed approximating set* A.

Let C be some set in $\mathbf{X}$. The task is to find the value

$$E(C, A, \mathbf{X}) = \sup_{x \in C} e(x, A, \mathbf{X}). \tag{3.6}$$

The value $E(C, A, \mathbf{X})$ is called *deviation* C from A. An element x_0 realizing $E(C, A, \mathbf{X})$, if it exists, has the worst of the best approximations among all $x \in C$.

Problem III *on the best approximation of assigned set* C *with a class* $\mathcal{A}$ *of approximating sets* A.

Let $\mathcal{A}$ be some class of approximating sets A. The task is to determine a value

$$\mathcal{E}(C, \mathcal{A}, \mathbf{X}) = \inf_{A \in \mathcal{A}} E(C, A, \mathbf{X}). \tag{3.7}$$

A set A_0 (if it exists) for which

$$\mathcal{E}(C, \mathcal{A}, \mathbf{X}) = E(C, A_0, \mathbf{X}) \tag{3.8}$$

is such that for the sets C and A_0 the worst the best approximation is the least one.

The following classes of approximating sets are the most frequently used.

(1) Class $E_N \subset \mathbf{X}$ of all linear spaces[1] of dimension $m \leq N$. Here the value $\mathcal{E}(C, E_N, \mathbf{X})$ is denoted through $d_N(C, \mathbf{X})$ and called the *Kolmogorov N-width*;

(2) Class $\mathbf{M}_N \subset \mathbf{X}$ of all point sets where the number of elements does not exceed N. The value $\mathcal{E}(C, M_N, \mathbf{X})$ is denoted through

[1] A definition of linear spaces is given in Section 5.

$\varepsilon_N(C, \mathbf{X})$ and its inverse function through $\mathcal{N}^0_\varepsilon(C, \mathbf{X})$; then [see (2.3)] the value

$$H_\varepsilon(C, \mathbf{X}) = \log_2 \mathcal{N}^0_\varepsilon(C, \mathbf{X}) \tag{3.9}$$

is called ε-*entropy (as related to* $\mathbf{X}$) *of the set* C.

If $N = 1$ and $x_0 \in M_1$ realizes $\varepsilon_1(C, \mathbf{X})$, then the element is called the *Chebyshev center of the set* C.

By the Hausdorff theorem $\varepsilon_N(C, \mathbf{X}) \to 0$ for $N \to \infty$ iff the set C is compact.

Problem IV *on the best approximation of an assigned set with the fixed approximation method.*

Let C and A be subsets of $\mathbf{X}$, and let P be an operator: $C \to A$. The task is to find a value

$$G(C, P, \mathbf{X}) = \sup_{x \in C} \rho(x, Px). \tag{3.10}$$

Let us show six remarkable approximation methods.

Example 3.4. The Fourier method where a function

$$x(t) = \sum_{k=0}^{\infty} (a_k \cos kt + b_k \sin kt) \tag{3.11}$$

is compared to trigonometrical polynomial $S_n(t)$ [see (3.4) and (3.5)].

Example 3.5. The Fejér method comparing the functions (3.11) and (3.5) to the trigonometrical polynomial

$$\sigma_n(t) = \frac{1}{n+1} \sum_{k=0}^{n} S_k(t)$$

$$= \sum_{k=0}^{n} \left(1 - \frac{k}{n+1}\right) (a_k \cos kt + b_k \sin kt). \tag{3.12}$$

Example 3.6. The analytical in the circle $|z| < a$ function

$$f(z) = \sum_{k=0}^{\infty} a_k z^k$$

is compared to a polynomial

$$P_n(z) = \sum_{k=0}^{n} a_k z^k. \tag{3.13}$$

The so-called "telescopic" method lowering the order of $P_n(x)$ is presented in Section 10 of Chapter 1.

Example 3.7. Consider the problem of approximation of a subset of a function from $C([-1,1],w)$ which is characterized with the condition

$$\max_{-1\le x\le 1} |f^{(n)}(x)| \le R, \qquad R > 0, \tag{3.14}$$

by the Lagrange interpolation polynomials with n knots. Let $-1 \le x_1 < x_2 < \ldots < x_n \le 1$

$$l_k(x) = \prod_{i\neq k}(x - x_i) / \prod_{i\neq k}(x_k - x_i).$$

It is known that if a function $f(x) \in C^n[-1,1]$, then

$$f(x) = L_{n-1}(x) + R_{n-1}(x), \qquad L_{n-1}(x_k) = f(x_k), \qquad k = 1, \ldots, n,$$

where a polynomial of the $(n-1)$th power (called the *Lagrange polynomial*) has the form

$$L_{n-1}(x) = \sum_{k=1}^{n} f(x_k) l_k(x), \tag{3.15}$$

with the method's error which appears while substituting the Lagrange interpolation polynomial on [-1,1] for the function $f(x)$:

$$R_{n-1}(x) = \frac{f^{(n)}(\xi)}{n!}\omega_n(x), \tag{3.16}$$

where

$$\omega_n(x) = \prod_{i=1}^{n}(x - x_i) \tag{3.17}$$

and the values of $\xi = \xi(x)$ belong to [-1,1]. Therefore, with regard to (3.16), (3.17), and the estimate (3.14), on class C, which is a set of functions from the space $C([-1,1],w)$ satisfying conditions (3.14) for the operator P which compares the function $f(x)$ to its interpolation polynomial $L_{n-1}(x)$, we have

$$G(C,P,C([-1,1],w)) = \sup_{f\in C} \sup_{-1\le x\le 1} \left| \frac{f^{(n)}(\xi)}{n!} \omega_n(x) w(x) \right|$$

$$= \frac{1}{n!} \sup_{f\in C} \max_{-1\le x\le 1} |f^{(n)}(x)| \max_{-1\le x\le 1} |\omega_n(x)w(x)|$$

$$= \frac{R}{n!} \max_{-1\le x\le 1} |\omega_n(x)w(x)|. \quad (3.18)$$

Example 3.8. Let $f(x) \in C[0,1]$. A polynomial

$$B_n(x) = \sum_{k=0}^{n} f\left(\frac{k}{n}\right) C_n^k x^k (1-x)^{n-k} \quad (3.19)$$

is called the *Bernstein polynomial.*

Example 3.9. The de la Vallée-Poussin method comparing functions (3.11) and (3.5) to a trigonometrical polynomial of power $2n-1$ is

$$v_{2n-1}(t) = \frac{1}{n} \sum_{k=n}^{2n-1} S_k(t). \quad (3.20)$$

Problem V *on the best approximation method.*

Let $\mathcal{P} = \{P\}$ some aggregate of operators $P: C \to A$. The task is to find the value

$$S(C,\mathcal{P},\mathbf{X}) = \inf_{P\in\mathcal{P}} G(C,P,\mathbf{X}). \quad (3.21)$$

§ 4. The Contraction Mapping Principle

Fixed points of operator. Contraction operators. Contraction mapping principle and method of successive approximations. Approximations for solution of nonlinear equations, systems of nonlinear algebraic equations, integral and ordinary differential equations.

Banach and Caccopoly formulated the contraction mapping principle broadly used in many divisions of mathematics. This principle underlies the construction of converging iteration methods for the solution of operator equations (see Chapter 3).

1. The Contraction Mapping Principle

Let $\mathbf{X}$ be a complete metric space with a distance $\rho(x, y)$, $x, y \in \mathbf{X}$ and let Ω be a closed set in $\mathbf{X}$. Let an operator P be assigned on Ω translating Ω into itself. We call an element $u \in \Omega$ the *fixed point* of the operator P if

$$u = P(u). \tag{4.1}$$

So, fixed points are a solution to Equation (4.1).

An operator P is called the *contraction* (*contraction operator*) on Ω if for $\forall\ x, y \in \Omega$ the condition

$$\rho(P(x), P(y)) \leq \alpha\rho(x, y) \tag{4.2}$$

is satisfied where $0 \leq \alpha < 1$ does not depend on x, y. Let a sequence $u^k \in \Omega$, $k = 0, 1, \ldots$ be such that

$$u^{k+1} = P(u^k). \tag{4.3}$$

Then we say that the operator P assignes on Ω the *iteration method* or the method of *successive approximations*; the recursive sequence $\{u^k\}$ is referred to as *iterative*. The following main theorem (Figure 3) is valid.

Theorem 1 (*the contraction mapping principle*). *If P is a contraction on Ω, then there exists on Ω a unique solution u of Equation* (4.1) *which can be obtained as a limit of the sequence* (4.3) *where $u^0 \in \Omega$ is an arbitrary element. A rate of convergence of the sequence $\{u^k\}$ to the solution is estimated by the inequality*

$$\rho(u^k, u) \leq \alpha^k(1 - \alpha)^{-1}\rho(u^1, u^0), \qquad k = 1, 2, \ldots . \tag{4.4}$$

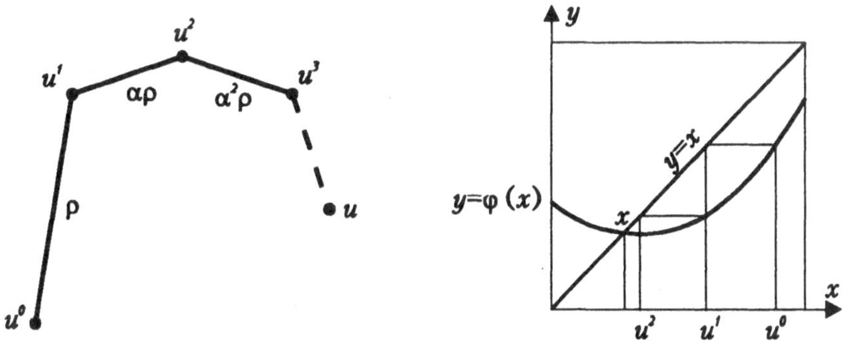

Figure 3

□ Since for $k \geq 1$ $u^{k+1} = P(u^k)$, $u^k = P(u^{k-1})$ we have, according to (4.2), that $\rho(u^{k+1}, u^k) \leq \alpha\rho(u^k, u^{k-1}) \leq \ldots \leq \alpha^k \rho(u^1, u^0)$. Therefore, for $p > 0$ by virtue of the triangle inequality

$$\rho(u^{k+p}, u^k) \leq \rho(u^{k+p}, u^{k+p-1}) + \rho(u^{k+p-1}, u^{k+p-2}) + \ldots$$

$$+\rho(u^{k+1}, u^k) \leq (\alpha^{k+p-1} + \alpha^{k+p-2} + \ldots + \alpha^k)\, \rho(u^1, u^0)$$

$$\leq \alpha^k \sum_{i=0}^{\infty} \alpha^i \rho(u^1, u^0) = \alpha^k (1-\alpha)^{-1} \rho(u^1, u^0). \quad (4.5)$$

Inequality (4.5) shows that the sequence $\{u^k\}$ is fundamental (in fact, $\alpha^k \to 0$) and due to completeness of the space $\mathbf{X}$ it converges to some element $u \in \mathbf{X}$. But, $u^k \in \Omega$, therefore, by virtue of the closeness of Ω, $u \in \Omega$ and the operator $P(u)$ has a sense. By inequality (4.2)

$$0 \leq \rho(u^{k+1}, P(u)) = \rho(P(u^k), P(u)) \leq \alpha\rho(u^k, u)$$

but $\rho(u^k, u) \to 0$ for $k \to \infty$, consequently, $u^{k+1} \to P(u)$; that is, relationship (4.1) is satisfied which means that u is a solution to this equation.

Uniqueness of the solution follows from inequality (4.2). In fact, if x were a second solution to (4.1) and $u \neq x$, then

$$\rho(x, u) = \rho(P(x), P(u)) \leq \alpha\rho(x, u)$$

which can only be in the case of $\rho(x, u) = 0$, that is, when $u = x$. The contradiction we have arrived at proves the uniqueness. Estimate (4.4) follows from (4.5) by passing to a limit in this inequality for $p \to \infty$. This estimate determines the domain of location for exact solution to equation (4.1). For example, for $k = 1$ we obtain

$$\rho(u, u^1) \leq \alpha(1-\alpha)^{-1} \rho(u^1, u^0). \blacksquare \quad (4.6)$$

Condition (4.2) as formulated in this theorem cannot, generally speaking, be replaced with weaker condition

$$\rho(P(x), P(y)) < \rho(x, y) \qquad \forall\, x, y \in \Omega, \qquad x \neq y. \quad (4.7)$$

The following example confirms this.

Example 4.1. Let $\mathbf{X} = \Omega = \mathbf{R}^1$ be a set of all real numbers, $\rho(x, y) = |x - y|$ and let $P(x) = |x| + (1 + |x|)^{-1}$. Then

$$|P(x) - P(y)| = [1 - (1 + |x|)^{-1}(1 + |y|)^{-1}]||x| - |y|| < |x - y|.$$

However, the mapping P has no fixed points.

The application of the contraction mapping principle to a concrete form of operator Equation (4.1) can lead to a difficulty in the following question: how to determine a size of closed space Ω that is mapped by the operator P onto itself and on which the operator P is a contraction operator. Let us present an algorithm guaranteeing (determining sufficient conditions) the fulfillment of the contraction mapping principle. Let a value r be determined by the equality

$$r = (1 - \alpha)^{-1}\rho(Pu^0, u^0). \tag{4.8}$$

Then it follows from inequality (4.5) for $k = 0$ that for $\forall\, p \geq 1$

$$\rho(u^p, u^0) \leq r.$$

Directing p to infinity ($u^p \to u$) in conditions of convergence we arrive at

$$\rho(u, u^0) \leq r.$$

We see thus that all successive approximations obtained with formula (4.3) and exact solution of Equation (4.1) belong to closed ball $\overline{D}_r(u^0)$. All conclusions and layings out in the proof of Theorem 1 stay true and condition (4.2) is satisfied for $\forall\ x, y \in \overline{D}_r(u^0)$; consequently, the results formulated in the theorem are true too. It means that the following theorem is valid.

Theorem 2. *If an operator P is defined on some part Ω of the complete metric space $\mathbf{X}$ and contains closed ball $\overline{D}_r(u^0)$ where r is determined by equality* (4.8) *and at the same time for $\forall\ x, y \in \overline{D}_r(u^0)$ condition* (4.2) *is satisfied, then there exists a unique solution u of Equation* (4.1) *$u \in \overline{D}_r(u^0)$ and it can be obtained as a limit of sequence* (4.3) *with initial approximation u^0; in addition*

$$\rho(u^k, u) \leq \alpha^k r. \tag{4.9}$$

2. Examples of the application of the contraction mapping principle

Example 4.2. *Algebraic and transcendental equations.* We need to find a root of the vequation

$$f(x) = 0 \tag{4.10}$$

under definite constraints superimposed on the scalar function $f(x)$ of one variable x. These constraints are determined in the course of the reasoning to follow. Let Equation (4.10) be transformed somehow to the form

$$x = \varphi(x) \tag{4.11}$$

in order to perform iterations (4.3) with the formula (see Figure 3)

$$u^{k+1} = \varphi(u^k). \tag{4.12}$$

Let it be a transformation such that all the roots of Equations (4.10) and (4.11) coincide. Notice that a type of transformation of Equation (4.10) to the type of Equation (4.11) may, generally speaking, affect essentially the characteristics related to the convergence rate of iteration process (4.12) and to its application.

Here is a simple transformation

$$x = x - \alpha(x) f(x),$$

where $\alpha(x)$ is some continuous function of fixed sign. In the simplest case $\alpha =$ const. If $f(x)$ is a differentiable function and $f'(x) \neq 0$, it is possible to take $\alpha(x) = (f'(x))^{-1}$ and then iteration process (4.12) converts into the classical *Newton method* for finding a root of Equation (4.10) (see Section 8 of Chapter 3)

$$u^{k+1} = u^k - (f'(u^k))^{-1} f(u^k). \tag{4.12$'$}$$

There are certain standard types of transformations and each of them has a certain name for the iteration process (4.12).

So, applying some transformation we have rewritten Equation (4.10) into the form (4.11). Let a function $\varphi(x)$ satisfy in some circle $D = \{|x - x_0| \leq r\}$ of the plane of the complex variable, the Lipschitz condition

$$|\varphi(x') - \varphi(x'')| \leq \alpha |x' - x''| \tag{4.13}$$

for $\forall\, x', x'' \in D$ and let the inequality

$$|x_0 - \varphi(x_0)| \leq (1-\alpha)r$$

be satisfied in the point $0 \leq \alpha < 1$ x_0.

Then the conditions of Theorem 2 are satisfied and the iteration process (4.12) started with $u^0 = x_0$ converges and the convergence rate is estimated with inequality (4.9).

Example 4.3. *Systems of algebraic and transcendental equations.* Let a system of n equations be reduced by some transformation to the equivalent form

$$x_i = \varphi_i(x_1, x_2, \ldots, x_n), \qquad i = 1, 2, \ldots, n. \tag{4.14}$$

Let $x = (x_1, x_2, \ldots, x_n)$; then $x \in \mathbf{R}^n$. Determine a metric in $\mathbf{R}^n$ with formula (1.5). Then a distance between elements x', $x'' \in \mathbf{R}^n$ is determined with the formula

$$\rho(x', x'') = \max_i |x_i' - x_i''|.$$

Let us formulate sufficient convergence conditions to a solution of system (4.14) for the iteration process

$$u_i^{k+1} = \varphi_i(u_1^k, u_2^k, \ldots, u_n^k), \qquad i = 1, 2, \ldots, n, \tag{4.15}$$

$k = 0, 1, \ldots$, $u_i^0 = x_i^0$, where $x^0 = (x_1^0, x_2^0, \ldots, x_n^0)$ is some vector of initial approximation.

Let in some ball $\overline{D}_r(x_0)$ (as a matter of fact, with our metric this ball is a cube in the space $\mathbf{R}^n$) the system of functions $\varphi_i(x)$, $i = 1, 2, \ldots, n$ satisfy the Lipschitz condition with a constant α $(0 \leq \alpha < 1)$

$$\max_i |\varphi_i(x') - \varphi_i(x'')| \leq \alpha\rho(x', x'') \tag{4.16}$$

for $\forall\ x', x'' \in \overline{D}_r(x_0)$ and let the inequality

$$\rho(x_0, \varphi(x_0)) \leq (1-\alpha)r \tag{4.17}$$

be valid in the point x_0 where $\varphi = (\varphi_1, \varphi_2, \ldots, \varphi_n)$. Statements of Theorem 2 are valid under these conditions.

Example 4.4. *The solution of a system of linear algebraic equations with an iteration method.* Take a special case of a system of equations of the type (4.14):

$$x_i = \sum_{j=1}^{n} a_{ij}x_j + b_i, \qquad i = 1, 2, \ldots, n \tag{4.18}$$

with a matrix $A = \{a_{ij}\}$; here $Px = Ax + b$. Iteration method (4.15) will be written as

$$u_i^{k+1} = \sum_{j=1}^{n} a_{ij} u_j^k + b_i$$
$$i = 1, 2, \ldots, n, \qquad k = 0, 1, \ldots, \qquad u_i^0 = x_i^0. \tag{4.19}$$

Since

$$\max_i \left| \sum_{j=1}^{n} a_{ij}(x_j' - x_j'') \right| \leq \max_i \sum_{j=1}^{n} |a_{ij}||x_j' - x_j''|$$

$$\leq \max_j |x_j' - x_j''| \cdot \max_i \sum_{j=1}^{n} |a_{ij}| = \rho(x', x'') \max_i \sum_{j=1}^{n} |a_{ij}|$$

inequality (4.16) will be satisfied with the constant

$$\alpha = \max_i \sum_{j=1}^{n} |a_{ij}| \tag{4.20}$$

on all space $\mathbf{R}^n$; therefore, we can assume $r = \infty$ in inequality (4.17) if $0 \leq \alpha < 1$. Consequently, for $0 \leq \alpha < 1$ Theorem 1 is valid, which in this case is appropriate in equivalent formulation.

Theorem 3. If matrix A of system (4.18) *is such that* $0 \leq \alpha < 1$ *where the value α is determined with formula* (4.20), *then the system of equations* (4.18) *has a unique solution. The solution can be obtained with iteration method* (4.19) *starting with arbitrary initial vector $x^0 = (x_1^0, x_2^0, \ldots, x_n^0)$. The degree of convergence for iterations is estimated with inequality* (4.4). *The condition*

$$\max_i \sum_{j=1}^{n} |a_{ij}| < 1 \tag{4.21}$$

is sufficient for the convergence of iterations.

Introducing another metric in $\mathbf{R}^n$ we get another sufficient condition for the convergence. Let a metric be assigned in $\mathbf{R}^n$ with formula (1.4); then

$$\rho(Ax', Ax'') = \sqrt{\sum_{i=1}^{n}\left[\sum_{j=1}^{n} a_{ij}(x'_j - x''_j)\right]^2}$$

$$\leq \sqrt{\sum_{i=1}^{n}\left[\sum_{j=1}^{n} a_{ij}^2 \sum_{j=1}^{n}(x'_j - x''_j)^2\right]} = \rho(x', x'')\sqrt{\sum_{i=1}^{n}\sum_{j=1}^{n} a_{ij}^2};$$

therefore, the following inequality is a sufficient condition for the convergence of the method of successive approximations in this case

$$\alpha = \left(\sum_{i=1}^{n}\sum_{j=1}^{n} a_{ij}^2\right)^{1/2} < 1. \tag{4.22}$$

Finally, if we introduce a metric in $\mathbf{R}^n$ with formula (1.3) then, after the estimation $\rho(Ax', Ax'')$ in this metric, we obtain the third sufficient condition for the convergence of the iteration method:

$$\alpha = \max_j \sum_{i=1}^{n} |a_{ij}| < 1. \tag{4.23}$$

Later (see Section 4 of Chapter 2 and Section 1 of Chapter 3) we find weaker conditions for the convergence of the method (4.18) (for instance, when all eigenvalues of A are strictly inside a unit circle).

Example 4.5. *Fredholm Integral Equations of the Second Kind.* Let $K(t, s)$ be a real continuous function called a kernel, defined in a square $a \leq t \leq b$, $a \leq s \leq b$, and let a function $f(t) \in C[a, b]$. Show that then the integral equation

$$x(t) = \lambda \int_a^b K(t, s)x(s)ds + f(t) \tag{4.24}$$

has a unique solution $x(t) \in L_2(a, b)$ for a sufficiently small absolute value of a parameter λ while employing an iteration method of type

$$u^{k+1}(t) = \lambda \int_a^b K(t, s)u^k(s)ds + f(t). \tag{4.25}$$

Let us denote

$$\int_a^b \int_a^b K^2(t,s)dtds = M^2 < \infty. \tag{4.26}$$

Take an operator $Ax = \lambda \int_a^b K(t,s)x(s)ds + f(t)$. Let us show that it translates each function $x(t) \in L_2(a,b)$ [the metric is defined in $L_2(a,b)$ by formula (1.15)] again into a function in $L_2(a,b)$. It is sufficient to prove, since $f(t) \in L_2(a,b)$, that the operator

$$y(t) = A_0 x = \int_a^b K(t,s)x(s)ds \tag{4.27}$$

translates each function $x(t) \in L_2(a,b)$ into a function from the same space. Apply the Cauchy–Bunyakovskii to obtain

$$y^2(t) = \left(\int_a^b K(t,s)x(s)ds\right)^2 \leq \int_a^b K^2(t,s)ds \int_a^b x^2(s)ds;$$

that is,

$$\int_a^b y^2(t)dt \leq M^2 \int_a^b x^2(s)ds. \tag{4.28}$$

Now estimate $\rho(Ax, Ay)$. We have

$$\rho(Ax,Ay) = \left[\int_a^b \left(\lambda \int_a^b K(t,s)x(s)ds - \lambda \int_a^b K(t,s)y(s)ds\right)^2 dt\right]^{1/2}$$

$$= |\lambda| \left[\int_a^b \left(\int_a^b K(t,s)(x(s)-y(s))ds\right)^2 dt\right]^{1/2}$$

$$\leq |\lambda| \left[\int_a^b \int_a^b K^2(t,s)ds\,dt\right]^{1/2} \left[\int_a^b (x(s)-y(s))^2 ds\right]^{1/2} \leq |\lambda| M \rho(x,y).$$

Consequently, if

$$\alpha = |\lambda| \cdot M < 1 \tag{4.29}$$

then applicability conditions for Theorem 1 are satisfied. So, existence and uniqueness of a solution of the preceding integral equation and convergence of iteration method (4.25) are proved for values of the parameter λ satisfying inequality (4.29). We continue in Section 13 of Chapter 2 the investigation of properties of operators A_0 of type (4.27) and of solutions of integral equations of type (4.24).

Example 4.6. *Ordinary differential equations.* The theorems on existence and uniqueness of a solution are proved in the courses on ordinary differential equations for the Cauchy problem

$$y'(x) = f(y,x), \qquad y(x_0) = y_0, \qquad x_0 \le x \le x_1 \tag{4.30}$$

by transforming differential Equation (4.30) into the equivalent integral equation with variable upper limit

$$y(x) = y_0 + \int_{x_0}^{x} f(y(t),t)\,dt \tag{4.31}$$

and then applying the method of successive approximations

$$u^{k+1}(x) = y_0 + \int_{x_0}^{x} f(u^k(t),t)\,dt \tag{4.32}$$

which is called the *Picard method* in this case after one of the founders of iteration methods. Let the function $f(y,x)$ be for $\forall\, y$ and $x \in [x_0, x_1]$ continuous bounded function satisfying with respect to y the Lipschitz condition with a constant L:

$$|f(y_2,x) - f(y_1,x)| \le L|y_2 - y_1|, \qquad x \in [x_0, x_1].$$

Let $\delta = x_1 - x_0$; then for $x \in [x_0, x_1]$ we obtain by (4.32)

$$|u^{k+1} - u^k| = |\int_{x_0}^{x} (f(u^k,t) - f(u^{k-1},t))dt| \le L\delta|u^k - u^{k-1}|.$$

Therefore, for $L\delta < 1$ the convergence of the Picard method follows from the contraction mapping principle. Algorithm (4.32) is rather laborious as applied to the solution of the Cauchy problem, and there are simpler approximate methods to find a solution to the Cauchy problem (see Section 11 of Chapter 3).

§ 5. Linear Spaces

Axioms and properties. Examples. Subspace. Dimension and linear dependence. Linear hull, a basis. Linear mapping, kernel of mapping, lemma on one-to-one mapping. Space of linear mappings $\mathcal{L}(\mathbf{X}, \mathbf{Y})$. Isomorphism of linear spaces. Convex sets.

1. Definitions, axioms, simple properties

A set E of elements $x, y, z, \ldots$ is called a *linear space* if the following operations are defined in it.

(1) Each two elements $x, y \in E$ have a definite corresponding element $x + y \in E$ called their *sum*;

(2) each element $x \in E$ and each number (scalar) λ have a definite corresponding element $\lambda x \in E$, *product* of element x by the *scalar* λ so that the following properties (axioms) are valid for $\forall\ x, y, z \in E$ and any scalars λ, μ.

(a) $x + y = y + x$, the commutativity law;

(b) $x + (y + z) = (x + y) + z$, the associativity law;

(c) there exists an element $0 \in E$ such that $x + 0 = x$;

(d) $\lambda(\mu x) = (\lambda\mu)x$;

(e) $1 \cdot x = x,\ 0 \cdot x = 0$ (on the left 0 is a scalar, on the right it is an element of the set E);

(f) $\lambda(x + y) = \lambda x + \lambda y$;

(g) $(\lambda + \mu)x = \lambda x + \mu x$.

Notice that the nature of elements $x, y, \ldots$ does not matter at all; it makes no difference how the sum $x + y$ and the product λx are defined. It is just required that these notions satisfy all axioms. Therefore, each time, when we encounter two operations satisfying the preceding conditions, we have a right to regard them as operations of addition and multiplication by a number.

Real or complex numbers serve as scalars $\lambda, \mu, \ldots$ in linear space. In the first case E is referred to as *real* (actual), in the second as *complex* linear space.

It is possible to define in linear space E for $\forall\ x \in E$ the *opposite* element $-x$ and, therefore, the subtraction operation of elements $y - x$, assigning by definition $-x = (-1)x$. Then, according to Axioms (e) and (g), $x + (-x) = 1 \cdot x + (-1)x = 0 \cdot x = 0$. We understand an expression $x - y = x + (-y)$ as the *difference* $x - y$. Let us present some simple facts.

Corollary 1. *The zero element is unique.*

□ Let 0_1 and 0_2 be zeros in E; then $0_1 + 0_2 = 0_1$, $0_2 + 0_1 = 0_2$ [according to the Axiom (c)] whence, by Axiom (a), $0_1 = 0_2$. ■

Corollary 2. *If* $\lambda x = \mu x$ *where* $x \neq 0$, *then* $\lambda = \mu$; *if* $\lambda x = \lambda y$ *and* $\lambda \neq 0$, *then* $x = y$.

□ Add $-\mu x$ to both sides of the equality $\lambda x = \mu x$ to obtain $(\lambda - \mu)x = 0$. If $\lambda \neq \mu$, then by Axiom (d) $(\lambda - \mu)^{-1}[(\lambda - \mu)x] = x = 0$; the contradiction so obtained confirms the first statement of the corollary. Adding $-\lambda y$ to both sides of the equality $\lambda x = \lambda y$ we obtain $\lambda(x - y) = 0$. Consequently, for $\lambda \neq 0$ we have $x - y = \lambda^{-1}[\lambda(x - y)] = 0$; that is, $x = y$. ■

2. Examples of Linear Spaces

Metric spaces $\mathbf{R}^1$, $\mathbf{R}^n$, $C[a,b]$, $C(\bar{\Omega})$, $C^k[a,b]$, $C^k(\bar{\Omega})$, $L_1(\Omega)$, $L_2(\Omega)$, $L_p(\Omega)$, m, l_1, and l_2, as we have discussed before, are at the same time linear spaces with the natural operation of adding of elements and the operation of multiplication of an element by a scalar. A space of all polynomials $P_n(x)$ of a power not exceeding n is also linear space. A set A_{nm} of all rectangular matrices of an order $m \times n$ with scalar elements, if the operations there for the matrices $A = \{a_{ij}\}$ and $B = \{b_{ij}\}$ are defined with formulas

$$(a_{ij}) + (b_{ij}) = (a_{ij} + b_{ij}), \qquad \lambda(a_{ij}) = (\lambda a_{ij})$$

is also linear space. A set of complex-valued solutions of an ordinary differential equation in partial derivatives composes complex linear space.

3. Subspace. Dimension and linear dependence. Linear hull, a basis

A set $E_1 \subset E$ is called a *subspace* of linear space E if $x, y \in E_1$ implies: for $\forall \lambda, \mu$ an element $\lambda x + \mu y \in E_1$ for $\forall x, y \subset E_1$. It is obvious that $0 \in E_1$.

If $x_1, x_2, \ldots, x_n \in E$, then any sum sort of $\sum_{k=1}^{n} \lambda_k x_k$ is called a *linear combination* of the elements $x_1, x_2, \ldots, x_n$. The finite system of elements $x_1, x_2, \ldots, x_n \in E_n$ $(x_k \neq 0)$ is referred to as *linearly dependent*, if there exist $\lambda_1, \lambda_2, \ldots, \lambda_n$ with $\sum_{k=1}^{n} |\lambda_k| > 0$ such that

$$\sum_k \lambda_k x_k = 0. \tag{5.1}$$

If equality (5.1) is satisfied just by the condition $\lambda_1 = \lambda_2 = \ldots = \lambda_n = 0$, then the elements $x_1, x_2, \ldots, x_n$ are referred to as *linearly independent*. The infinite system of elements $\{x_k\}_1^\infty$, $x_k \in E$ in a linear space E is referred to as *linearly independent*, if any of its finite subsystem $x_{k_1}, x_{k_2}, \ldots, x_{k_n}$ is linearly independent.

Let $\{x_k\} \in M$ where M is some set form E. An aggregate of all possible finite linear combinations of elements of this set $\sum \lambda_i x_{k_i}$ where $x_{k_i} \in M$ and λ_i are arbitrary scalars is called the *linear hull of the set M* or *linear manifold.*

If there exists in E a system of n linearly independent elements whose hull coincides with E, then linear space E is referred to as *n-dimensional* (*finite-dimensional*) and any linearly independent system of n elements is termed a *basis* of the space E. Linear space that is not finite-dimensional is referred to as *infinite-dimensional.*

Let $\{e_k\}_1^n$ be a basis in a n-dimensional linear space and let x be any element of E. Since the space is n-dimensional, the elements $e_1, e_2, \ldots, e_n, x$ are linearly dependent; that is, $\exists\ \lambda_1, \lambda_2, \ldots, \lambda_n, \lambda_{n+1}$ such that $\sum_{i=1}^{n} \lambda_i e_i + \lambda_{n+1} x = 0$. In addition, $\lambda_{n+1} \neq 0$ since otherwise the elements $e_1, e_2, \ldots, e_n$ would be linearly dependent. Therefore,

$$x = \sum_{i=1}^{n} \xi_i e_i, \tag{5.2}$$

where $\xi_i = -\lambda_i / \lambda_{n+1}$.

Representation (5.2) of an arbitrary element of n-dimensional space E is called an *expansion* of the element with respect to the basis $\{e_i\}$ and it is unique. The numbers ξ_i, $i = \overline{1,n}$ are called the *coordinates* of the vector x in the basis $\{e_i\}_1^n$.

Functional analysis deals mostly with infinitely dimensional linear spaces. These are the spaces $C[a,b]$, $C^k[a,b]$, $L_1(a,b)$, $L_2(a,b)$, $L_p(a,b)$ where a system x^n, $n = 0,1,2,\ldots$ is a linearly independent system of elements. The spaces m, l_1, l_2 are also infinite-dimensional where a system of infinite-dimensional vectors of the type $e_n = (0,0,\ldots,1,0,\ldots,0,\ldots)$ with the unit at the nth position is the linearly independent system; $C(\bar{\Omega})$, $C^k(\bar{\Omega})$, $L_1(\Omega)$, $L_2(\Omega)$, $L_p(\Omega)$ are infinite-dimensional spaces as well.

4. Linear mapping, kernel of mapping

A mapping f of linear space $\mathbf{X}$ onto linear space $\mathbf{Y}$ is called a *linear mapping* (*linear operator*), if for any $x, y \in \mathbf{X}$ and any scalars λ, μ

$$f(\lambda x + \mu y) = \lambda f(x) + \mu f(y). \tag{5.3}$$

Let us denote a set of linear mappings $f : \mathbf{X} \to \mathbf{Y}$ by $\mathcal{L}(\mathbf{X},\mathbf{Y})$; if the addition of its elements and the multiplication by numbers are defined in a natural way:

$$(f_1 + f_2)(x) = f_1(x) + f_2(x) \quad \text{and} \quad (\lambda f)(x) = \lambda f(x)$$

it forms a linear space too. If $f : \mathbf{X} \to \mathbf{Y}$ is a linear mapping, then a set $\{x \in \mathbf{X} : f(x) = 0\}$ is termed a *kernel* of the mapping f and is denoted by $\ker f$:

$$\ker f = \{x : f(x) = 0\}. \tag{5.4}$$

Lemma 1. *It is necessary and sufficient for a linear mapping $f : \mathbf{X} \to \mathbf{Y}$ of linear space $\mathbf{X}$ into linear space $\mathbf{Y}$ to be a one-to-one mapping of $\mathbf{X}$ into $\mathbf{Y}$, that its kernel would consist only of zero element:* $\ker f = 0$.

□ *Necessity.* It is obvious that f translates a zero element in $\mathbf{X}$ into a zero element in $\mathbf{Y}$, since for $\forall\, x \in \mathbf{X}$, $f(0) = f(0\cdot x) = 0\cdot f(x) = 0$. Therefore, by virtue of one-to-one bijectivity, there is no element $x \neq 0$ such that $f(x) = 0$; that is, $\ker f = 0$.

Sufficiency. Let $\ker f = 0$; however, $f(x) = f(y)$. Then $f(x - y) = f(x) - f(y) = 0$; that is, $x - y \in \ker f$, consequently $x - y = 0$; that is, $x = y$. ■

Two linear spaces **X** and **Y** are referred to as *isomorphic* if it is possible to establish between elements of these spaces the one-to-one bijectivity that preserves algebraic operations; that is, such that if $x \leftrightarrow y$, $x' \leftrightarrow y'$, then $x + x' \leftrightarrow y + y'$ and $\lambda x \leftrightarrow \lambda y$.

Example 5.1. A space of polynomials with real coefficients of a degree not higher than n is isomorphic to a real space $\mathbf{R}^{n+1}$, since each polynomial $P_n(x) = \sum_{k=0}^{n} a_k x^k$ has corresponding vector $(a_0, a_1, \ldots, a_n) \in \mathbf{R}^{n+1}$.

Let $x_1, x_2 \in E$. An aggregate of all the points of the type

$$x = (1-t)x_1 + tx_2, \quad t \in [0,1] \tag{5.5}$$

is called a *segment* connecting the points x_1, x_2. A set $M \subset E$ in a linear space E is referred to as *convex* if it follows for $\forall\, x_1, x_2 \in M$ that the segment (5.5) belongs to M. A set $K \subset E$ is called a *cone* if it follows from $x \in K$ that $\alpha x \in K$ for $\alpha > 0$.

§ 6. Normed and Banach Spaces

Normed, seminormed, strictly normed spaces. Axioms. Properties of norms, convergence. Examples. Equivalent norms. Banach spaces. Subspace. Series. Completeness of a system of elements. Basis.

One can manage to introduce a metric in some linear spaces assigning a norm of an element; the latter notion is equivalent to the notion of a length of a vector in conventional space. A metric $\rho(x,y)$ in a space **X** of such kind is invariant relative to a shear; that is,

$$\rho(x,y) = \rho(x+z,\, y+z), \quad \forall\, x, y, z \in \mathbf{X} \tag{6.0}$$

and **X** is a linear metric space.

1. Definitions, axioms, simple properties of norms. Convergence

Linear space **X** (real or complex) is called *normed space*, if a real function is defined on a set of its elements $x \in \mathbf{X}$ called a *norm* and denoted by $\|x\|_{\mathbf{X}}$ or, in short, $\|x\|$ and satisfying the following properties (axioms) for $\forall\, x, y \in \mathbf{X}$ and for a scalar λ.

(1) $\|x\| \geq 0, \ x \in \mathbf{X}$;

(2) $\|\lambda x\| = |\lambda| \, \|x\|$;

(3) $\|x + y\| \leq \|x\| + \|y\|$;

(4) if $\|x\| = 0$, then $x = 0$.

It follows from Property (2) that if $x = 0$, then $\|x\| = 0$.

□ $\|0\| = \|0 \cdot x\| = 0 \cdot \|x\| = 0$. ■

If a real function $\|x\|$, $x \in \mathbf{X}$ is defined on elements of linear space $\mathbf{X}$ satisfying just Properties (1)–(3), then the space $\mathbf{X}$ is referred to as *seminormed* and the function $\|x\|$ is called a *seminorm*.

Normed space is referred to as *strictly normed* if the equality sign in inequality (3) for $x \neq 0$, $y \neq 0$ is attained just for $y = \lambda x$ where $\lambda > 0$. It is possible to introduce a distance in a normed space between its elements with the formula

$$\rho(x, y) = \|x - y\| \tag{6.1}$$

or

$$\rho(x, y) = \alpha \|x - y\|,$$

where α is any positive fixed number. It means that any normed space is metric where a metric satisfies the relationship (6.0). Therefore, all the definitions we introduced while studying metric spaces (limit, ball, sphere, closed and open sets, etc. are valid for normed spaces and all theorems proved for metric spaces are also valid. The convergence in a metric space is a convergence with respect to a norm: $x_n \to x \iff \|x - x_n\| \to 0$ and inequality (1.1) transforms into the inequality

$$|\, \|x\| - \|y\| \,| \leq \|x - y\| \tag{6.2}$$

which implies that if $x_n \to x$, then $\|x_n\| \to \|x\|$; that is, the norm is a continuous function of x.

2. Examples

6.1. Space $\mathbf{R}^n$ is normed; it is possible to introduce norms in it, for instance, by formulas:

$$\|x\|_1 = \sum_{i=1}^{n} |x_i| \tag{6.3}$$

$$\|x\|_2 = \left(\sum_{i=1}^{n} x_i^2\right)^{1/2} \tag{6.4}$$

$$\|x\|_\infty = \max_i |x_i|. \tag{6.5}$$

6.2. Space l_1 normed with the norm

$$\|x\| = \sum_{i=1}^{\infty} |x_i|. \tag{6.6}$$

6.3. Space l_2 normed with the norm

$$\|x\| = \left(\sum_{i=1}^{\infty} x_i^2\right)^{1/2}. \tag{6.7}$$

6.4. Space m normed with the norm

$$\|x\| = \sup_i |x_i|. \tag{6.8}$$

6.5. Space $C[a,b]$ normed with the Chebyshev norm

$$\|f\| = \max_{x\in[a,b]} |f(x)|. \tag{6.9}$$

In a weight space $C([a,b],w)$ this norm is

$$\|f\| = \max_{x\in[a,b]} |f(x)w(x)| \tag{6.9'}$$

and in the space $C(\bar{\Omega})$

$$\|f\| = \max_{x\in\bar{\Omega}} |f(x)|. \tag{6.9''}$$

6.6. Space $C^k[a,b]$ normed with the norm

$$\|f\| = \sum_{i=0}^{k} \max_{x\in[a,b]} |f^{(i)}(x)| \tag{6.10}$$

in the space $C^k(\bar{\Omega})$

$$\|f\| = \sum_{0\le|\alpha|\le k} \max_{x\in\bar{\Omega}} |D^\alpha f|. \tag{6.10'}$$

One can introduce in the space $C^k[a,b]$ a norm with the formula

$$\|f\| = \max\left\{ \max_{x\in[a,b]} |f(x)|,\ \max_{x\in[a,b]} |f'(x)|, \dots, \max_{x\in[a,b]} |f^k(x)| \right\}. \tag{6.11}$$

If we take on elements of $C^k[a,b]$ the function

$$\|f\| = \max_{x\in[a,b]} |f^k(x)|, \qquad k \ge 1,$$

then this expression is a seminorm.

6.7. Spaces $L_1(a,b)$, $L_2(a,b)$, $L_p(a,b)$ are normed with the norms

$$\|f\|_{L_1} = \int_a^b |f(x)|dx \tag{6.12}$$

$$\|f\|_{L_2} = \left(\int_a^b |f(x)|^2 dx\right)^{1/2} \tag{6.13}$$

$$\|f\|_{L_p} = \left(\int_a^b |f(x)|^p\, dx\right)^{1/p}, \tag{6.14}$$

respectively. The spaces $L_1(\Omega)$, $L_2(\Omega)$, $L_p(\Omega)$ are normed in a similar way with the norms, respectively,

$$\|f\|_{L_1(\Omega)} = \int_\Omega |f(x)|d\Omega \tag{6.15}$$

$$\|f\|_{L_2(\Omega)} = \left(\int_\Omega |f(x)|^2 d\Omega\right)^{1/2} \tag{6.16}$$

$$\|f\|_{L_p(\Omega)} = \left(\int_\Omega |f(x)|^p d\Omega\right)^{1/p} \tag{6.17}$$

and in the weight space $L_p(\Omega, w)$

$$\|f\|_{L_p(\Omega,w)} = \left(\int_\Omega |f(x)|^p\, w(x)\, d\Omega.\right)^{1/p}. \tag{6.17'}$$

6.8. Space of polynomials $P_n(x) = \sum_{k=0}^{n} a_k x^k$ of a degree not higher than n is normed if a norm is introduced by formula

$$\|P_n\| = \sum_{k=0}^{n} |a_k|.$$

It is easy to check that the norms introduced in Examples 6.1–6.8 satisfy the preceding axioms.

The spaces $C[a,b]$, $C([a,b],w)$ are not strictly normed.

□ Let $a = 0$, $b = 1$ for the sake of simplicity. Take two functions f and $g \in C([0,1],w)$: $f(x) = x/w(x)$ and $g(x) = 1/w(x)$. For these functions $\|f\| = \|g\| = 1$, $\|f+g\| = 2$, altough there is no $\lambda > 0$ such that $f(x) = \lambda g(x)$. ■

3. Equivalent norms. Banach spaces. Subspace. Series

Two norms can be assigned on the same set of elements of linear space (see Examples 6.1 and 6.6). Two norms $\|x\|_1$ and $\|x\|_2$ in a normed space $\mathbf{X}$ are referred to as *equivalent*, if there exist the constants $c_1 > 0$ and $c_2 > 0$ such that for $\forall\ x \in \mathbf{X}$ the following inequality holds

$$c_1\|x\|_1 \le \|x\|_2 \le c_2\|x\|_1. \tag{6.18}$$

Obviously, if inequalities (6.18) are satisfied, then the inequalities

$$c_2^{-1}\|x\|_2 \le \|x\|_1 \le c_1^{-1}\|x\|_2 \tag{6.19}$$

are valid too. It follows from inequalities (6.18) and (6.19) that if $x_n \to x$ in some norm, the same will be true in another norm.

It is proven in the course of linear algebra that all norms in the space $\mathbf{R}^n$ as well as the norms defined with formulas (6.3)–(6.5) are equivalent. The equivalence is easily proved for the norms (6.10), (6.11) in the space $C^k[a,b]$, and for the norms (6.9), (6.9′) and (6.17), (6.17′) by $0 < C_1 \le W(k) \le C_2$.

Since normed spaces are metric, they can be complete and incomplete. Incomplete normed space may be completed in the same sense as the complement of metric spaces is understood. Complete normed space is called the *Banach space or B-space.*

Since the spaces $\mathbf{R}^n$, m, l_1, l_2, $C[a,b]$, $C^k[a,b]$, $L_1(a,b)$, $L_1(\Omega)$, $L_2(a,b)$, $L_2(\Omega)$, $L_p(a,b)$, $L_p(\Omega)$ are complete metric spaces, they are, being normed spaces, the Banach spaces as well. We assume hereinafter for the sake of simplicity that all normed spaces are complete, that is, Banach spaces, and denote them by the letter B.

Any linear closed set of a set B is called its *subspace*.

Since B-space is linear, linear combinations are defined there for a finite number of elements $\{x_i\}_1^n \in B$:

$$\sum_{i=1}^{n} \lambda_i x_i.$$

The introduction of a norm makes it possible to consider an infinite series in B-space composed of elements $\{x_n\} \in B$:

$$\sum_{k=1}^{\infty} x_k = x_1 + x_2 + \ldots + x_n + \ldots . \tag{6.20}$$

The series(6.20) is referred to as *convergent* if a sequence $\{S_n\}$ of its partial sums $S_n = \sum_{k=1}^{n} x_k$ converges. The limit of sequence $\{S_n\}$: $S = \lim_{n\to\infty} S_n$ is called a *sum of the series* (6.20). We say that series (6.20) *converges absolutely* if the following number series converges

$$\sum_{k=1}^{\infty} \|x_k\| = \|x_1\| + \|x_2\| + \ldots + \|x_n\| + \ldots . \tag{6.21}$$

Lemma 1. *In a B-space the absolute convergence implies conventional convergence; in addition, the estimate is valid:*

$$\left\| \sum_{k=1}^{\infty} x_k \right\| \leq \sum_{k=1}^{\infty} \|x_k\|.$$

□ In fact, if $m > n$, then $S_m - S_n = x_{n+1} + x_{n+2} + \ldots + x_m$ and, consequently, $\|S_m - S_n\| \leq \|x_{n+1}\| + \|x_{n+2}\| + \ldots + \|x_m\|$. Since the series (6.21) converges, the right-hand side of this inequality is arbitrarily small for $n \to \infty$. Therefore, the sequence $\{S_n\}$ is fundamental and, by virtue of completeness of B-space it converges to some element $S \in B$. We obtain the desired estimate by passing to a limit for $n \to \infty$ in the inequality $\|S_n\| \leq \sum_{k=1}^{n} \|x_k\|$ and making use of continuity of a norm. ■

4. Completeness of a system of elements. Basis

A system $\{x_\alpha\} \in B$, $\alpha \in U$ (U is some countable set of indices) is referred to as *complete in* B, if a set of all finite linear combinations of its elements (its linear hull) composes an everywhere dense in B set. A sequence of elements $\{e_n\} \in B$ is called a *basis*, if for $\forall\, x \in B$ there exists and, moreover, is unique a sequence of numbers $\{\lambda_n\}$, $n = 1, 2, \ldots$ such that

$$x = \sum_{n=1}^{\infty} \lambda_n e_n. \tag{6.22}$$

Formula (6.22) is called the *expansion* of an element $x \in B$ with respect to the basis $\{e_n\}$.

So, if $\{e_n\}$ is a basis and $x \in B$ is any element, then for $\forall\, \varepsilon > 0$ $\exists\, n_0$ such that for $n \geq n_0$

$$\|x - \sum_{k=1}^{n} \lambda_k e_k\| < \varepsilon;$$

that is, $\{e_n\}$ is complete in B system of elements.

Lemma 2. *If a system of elements $\{e_n\}$ composes a basis, then it is linearly independent.*

□ The statement of the lemma follows from the uniqueness of the expansion of elements x of the space with respect to the basis $\{e_n\}$. In fact, if $\{e_n\}$ happened to be a linearly dependent system, then a set $e_{n_1}, \ldots, e_{n_k}$ would be found such that the following equality is valid for the numbers $\lambda_1, \lambda_2, \ldots, \lambda_k$ $(\sum_{i=1}^{k} |\lambda_1| > 0)$: $0 = \sum_{i=1}^{k} \lambda_i e_{n_i}$. But the expansion $0 = \sum_{i=1}^{\infty} 0 \cdot e_i$ is obvious for a zero element. The contradiction we have arrived at proves the lemma. ■

Lemma 3. *If B-space has a basis, then it is separable.*

□ Let $\{e_n\}_1^\infty$ be a basis in B. Then a set of all linear combinations of the type $\sum_{i=n_1}^{n_2} r_i e_i (n_2 \geq n_1)$ with rational coefficients r_i is countable everywhere dense in B. ■

The inverse statement is incorrect: not every separable B-space has a basis.

Examples

6.9. An aggregate of elements $e_k = (\delta_{1k}, \delta_{2k}, \ldots, \delta_{kk}, \ldots)$ where δ_{ik} is the Kronecker symbol: $\delta_{ik} = 0$ for $i \neq k$ and $\delta_{ik} = 1$ for $i = k$, $k = 1, 2, \ldots$ composes a basis in the spaces l_1, l_2. Let us prove, for the sake of definiteness, this statement for the space l_2, since the proof for the space l_1 is performed in a similar way.

□ There is for $\forall\ x = (x_1, x_2, \ldots, x_i, \ldots)$ a unique representation $x = \sum_{i=1}^{\infty} x_i e_i$ since $\sum_{i=1}^{n} x_i e_i = (x_1, x_2, \ldots, x_n, 0, 0, \ldots)$ and, therefore,

$$\|x - \sum_{i=1}^{n} x_i e_i\| = \|(0, 0, \ldots, 0, x_{n+1}, x_{n+2}, \ldots)\| = \left(\sum_{i=n+1}^{\infty} |x_i|^2\right)^{1/2} \to 0$$

for $n \to \infty$; consequently,

$$x = \lim_{n\to\infty} \sum_{i=1}^{n} x_i e_i = \sum_{i=1}^{\infty} x_i e_i. \tag{6.23}$$

Then if $x = \sum_{i=1}^{\infty} x_i' e_i$ is another expansion for the element x, then

$$\sum_{i=1}^{\infty} (x_i - x_i') e_i = (x_1 - x_1',\ x_2 - x_2', \ldots) = 0;$$

that is, $x_i = x_i'$ $i = 1, 2, \ldots$ ■

6.10. It is possible to show that the space m is not separable; therefore, it has no basis. Limit transition is impossible for m in (6.23).

6.11. Let $B = C[0,1]$. Take in $C[0,1]$ a sequence of elements

$$x, 1 - x, u_{00}(x), u_{10}(x), u_{11}(x), u_{20}(x), u_{21}(x), u_{22}(x), \ldots \tag{6.24}$$

where $u_{kl}(x)$ ($k = 1, 2, \ldots,\ 0 \leq l < 2^k$) is defined as follows: $u_{kl}(x) = 0$ if $x \notin (l \cdot 2^{-k}, (l+1) \cdot 2^{-k})$ and inside this interval $u_{kl}(x)$ has a diagram in the form of an isosceles triangle whose height is equal to unit and whose base coincides with the segment $[l \cdot 2^{-k}, (l+1)2^{-k}]$ (Figure 4).

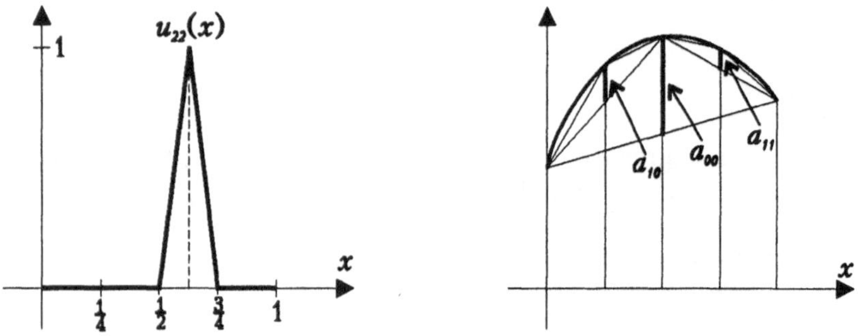

Figure 4

Any function $f(x) \in C[0,1]$ may be represented in the form of a series

$$f(x) = a_0 x + a_1(1-x) + \sum_{k=0}^{\infty} \sum_{l=0}^{2^k-1} a_{kl}\, u_{kl}(x),$$

where $a_0 = f(1), a_1 = f(0)$ and coefficients a_{kl} are found simply by geometric construction as shown in Figure 4. A diagram of the partial sum of a series

$$a_0 x + a_1(1-x) + \sum_{k=0}^{S-1} \sum_{l=0}^{2^k-1} a_{kl}\, u_{kl}(x)$$

is a broken line with $2^S + 1$ vertices lying on the curve $f(x)$ in the points with equidistant x-coordinates. Consequently, the aggregate of functions (6.24) composes a basis; these functions are the simplest so-called *spline functions.*

There exist linearly independent systems of elements such that if we remove finite and even countable number of their members, we obtain again a complete system. Let us investigate, as an example, the completeness in spaces $C[0,1]$, $L_2(0,1)$ of system x^n, where $n = 0, 1, \ldots$ Muntz had studied the question of for what integer nonnegative $n_0 < n_1 < n_2 \ldots$ the system of powers $x^{n_0}, x^{n_1}, x^{n_2}, \ldots$ is complete in $C[0,1]$, $L_2(0,1)$ and proved two theorems which we present here skipping the proofs.

Theorem 1. *It is necessary and sufficient for the system of powers* $\{x^{n_i}\}_0^{\infty}$ $(n_0 < n_1 < n_2 \ldots)$ *to be complete in* $C[0,1]$ *if:*

(1) $n_0 = 0$;

(2) *the series* $\sum_{i=1}^{\infty} 1/n_i$ *is nonconvergent.*

Theorem 2. *The system of powers* $\{x^{n_i}\}_0^\infty$ *is complete in* $L_2(0,1)$ *iff the series* $\sum_{i=1}^{\infty} 1/n$ *is nonconvergent.*

Take, in addition to these theorems, one more example. Let a function $f(x) \in C[1,2]$ and functions $f_1(x)$, $f_2(x) \in C[-2,2]$ be such that $f_1(x) \equiv f_2(x) \equiv f(x)$ for $x \in [1,2]$ and $f_1(x) = f_1(-x)$, $f_2(x) = -f_2(-x)$ for $x \in [-2,2]$. Then the function $f_1(x)$ can be as much as desired strictly approximated on [-2,2] with odd polynomials, and the function $f_2(x)$ with even ones. Consequently, any function $f(x)$ can be as much as desired strictly approximated on [1,2] with just even or just odd polynomials. Thus the linear hull of the system $\{x^n\}_0^\infty$ does not contract when even a countable number of its numbers are removed.

The latter property must lead (and does lead) to the unpleasant fact connected with the loss of stability for $n \to \infty$ while determining the coefficients λ_i^n of linear combinations $\sum_{i=0}^{n} \lambda_i^n x^i$ approximating some function $f(x)$ in the metric of spaces $C[0,1]$ or $L_2(0,1)$ that belongs to these spaces. This system of functions does not possess the so-called *minimality property*: removing a finite number of its numbers we obtain again a complete system of functions.

§ 7. Spaces with an Inner Product. Hilbert Spaces

Spaces with an inner product: Euclidean and unitary spaces; properties of inner product. Cauchy–Bunyakovskii inequality. Hilbert space and its properties. Strong and weak convergence. Parallelogram equality. Strict norming. Examples. Orthogonality and linear independence. Sonin–Schmidt process of orthogonalization. Orthogonal and orthonormed systems. Hermitian and symmetric bilinear forms. Formal construction of new (energetic) spaces. Sobolev W_2^1 space.

If we define, in addition, for the elements of linear normed space the concept of an angle between them, we get a space whose properties are close to those of our three-dimensional space.

1. Euclidean and unitary spaces. Axioms. Properties of inner product. Cauchy–Bunyakovskii inequality. Hilbert space and its properties. Strong and weak convergence. Parallelogram equality, strict norming

Real linear space E is referred to as *Euclidean* if each couple of its elements x, y has a real number (x, y) which is called an *inner product* so that the following axioms are valid.

(1) $(x, x) \geq 0$ and $(x, x) = 0$ iff $x = 0$;

(2) $(x, y) = (y, x)$;

(3) $(\lambda x, y) = \lambda(x, y)$ for any real λ;

(4) $(x + y, z) = (x, z) + (y, z)$.

We deduce from axioms (2), (3), and (4) that

$$(x, \lambda y) = \lambda(x, y), \qquad (x, y_1 + y_2) = (x, y_1) + (x, y_2). \tag{7.1}$$

A Euclidean space becomes a normed one if a norm there is defined with the formula

$$\|x\| = (x, x)^{1/2}. \tag{7.2}$$

Let us establish the following important *Cauchy–Bunyakovskii inequality*: for $\forall\, x, y \in E$

$$|(x, y)| \leq \|x\| \cdot \|y\|. \tag{7.3}$$

□ Since we have for any real λ by the axiom 1

$$(x - \lambda y, x - \lambda y) \geq 0$$

then, opening this expression, we obtain a square with respect to the λ nonnegative trinomial

$$\lambda^2(y, y) - 2\lambda(x, y) + (x, x) \geq 0$$

whose determinant is nonpositive; that is, $(x, y)^2 - \|x\|^2\|y\|^2 \leq 0$. Hence inequality (7.3) follows. ■

A norm defined by equality (7.2) satisfies all axioms of a normed space. In particular, it follows from Axiom (3) and from the first equality (7.1) that $\|\lambda x\| = |\lambda| \cdot \|x\|$ and the triangle inequality is verified as follows

$$\|x+y\|^2 = (x+y, x+y) = (x,x) + 2(x,y) + (y,y)$$

$$\leq \|x\|^2 + 2\|x\|\,\|y\| + \|y\|^2 = (\|x\| + \|y\|)^2.$$

A complex linear space E is called *unitary space* if each couple of its elements x, y has a complex-valued number (x, y) called an *inner product* so that Axioms (1), (3), and (4) are valid for any complex number λ as well as the axiom

$$(2^*) \quad (x,y) = \overline{(y,x)},$$

where the overbar means the conversion operation to a complex conjugate number.

Corollary 1. *For unitary space* $(x, \lambda y) = \bar{\lambda}(x, y)$.

□ Indeed, $(x, \lambda y) = \overline{\lambda(y,x)} = \bar{\lambda}\overline{(y,x)} = \bar{\lambda}(x,y)$. ■

A norm is introduced for unitary space by formula (7.2) and inequality (7.3) is valid.

Complete Euclidean space or complete unitary space is called *Hilbert space*; we denote it by the letter H. For the sake of simplicity, we study hereinafter just real separable Hilbert spaces.

Since a Hilbert space is linear normed Banach and, consequently, metric space at the same time, it satisfies all definitions, concepts, and statements we introduced while investigating metric linear normed and Banach spaces. The convergence with respect to a norm $x_n \to x$ for Hilbert spaces is called *strong convergence*.

Lemma 1. *An inner product* (x, y) *is a continuous function of its arguments*.

□ Let $x_n \to x$, $y_n \to y$ be any sequences. As we proved before (see Section 1), $\|y_n\| < C$ for $\forall\ n$ where $C > 0$ is some constant. Then using (7.3) we obtain

$$|(x_n, y_n) - (x,y)| = |(x_n - x, y_n) + (x, y_n - y)|$$

$$\leq \|x_n - x\|\,\|y_n\| + \|x\|\,\|y_n - y\| \to 0 \quad \text{for} \quad n \to \infty. \ ■$$

It is possible to introduce one more type of convergence in a Hilbert space along with strong convergence (the convergence with respect to the norm). A sequence $x_n \in H$ is referred to as *weakly convergent* to an element $x \in H$ ($x_n \rightharpoonup x$) if $\lim_{n\to\infty}(x_n, f) = (x, f)$ for $\forall\, f \in H$.

Lemma 2. *A sequence x_n cannot converge weakly to different elements of H.*

□ Let there exist two elements $x, \bar{x} \in H$ for which for $\forall\, f \in H$ $(x_n, f) \to (x, f)$ and $(x_n, f) \to (\bar{x}, f)$ for $n \to \infty$. Then $(x - \bar{x}, f) = 0$ for $\forall\, f \in H$; in particular, for $f = x - \bar{x}$ we have $(x - \bar{x},\, x - \bar{x}) = 0$; that is, $x = \bar{x}$. ■

Lemma 3. *If a sequence x_n converges strongly to x, then it converges weakly to x.*

□ Indeed,

$$|(x_n, f) - (x, f)| = |(x_n - x, f)| \le \|x_n - x\|\,\|f\| \to 0$$

for $n \to \infty$. ■

Lemma 4. *If a sequence x_n converges weakly, it is bounded.*

□ Lemma 4 follows by the Banach–Steinhaus theorem for linear functionals (Chapter 2, Section 6). ■

The so-called *parallelogram equality* is valid in Hilbert spaces (Figure 5):

$$\|x + y\|^2 + \|x - y\|^2 = 2(\|x\|^2 + \|y\|^2) \tag{7.4}$$

$$\Box\, \|x + y\|^2 + \|x - y\|^2 = (x + y, x + y) + (x - y, x - y)$$

$$= (x, x) + 2(x, y) + (y, y) + (x, x) - 2(x, y) + (y, y) = 2(\|x\|^2 + \|y\|^2).\ \blacksquare$$

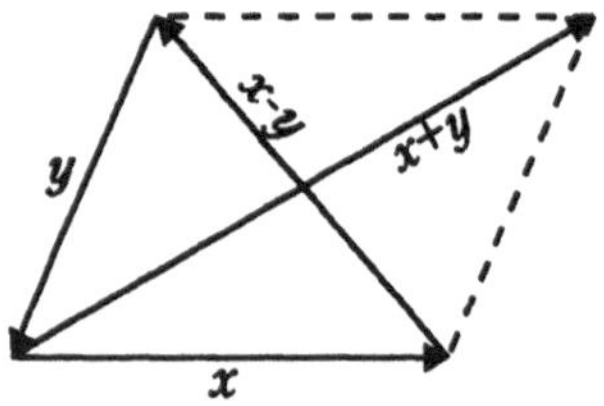

Figure 5

Computing $\|x + y\|^2 - \|x - y\|^2$ we obtain in a similar way the following connection between a norm and inner product in real Hilbert space:

$$4(x, y) = \|x + y\|^2 - \|x - y\|^2 \tag{7.5}$$

or computing $||x+y||^2 - ||x||^2 - ||y||^2$:

$$2(x,y) = ||x+y||^2 - ||x||^2 - ||y||^2.$$

Lemma 5. *Hilbert space is a strictly normed space.*

□ Let the elements $x, y \in H$ be such that $||x+y|| = ||x|| + ||y||$. Square this equality to obtain

$$||x||^2 + 2(x,y) + ||y||^2 = ||x||^2 + 2||x||\,||y|| + ||y||^2;$$

that is, $(x,y) = ||x||\,||y|| > 0$. Then

$$||x-\alpha y||^2 = (x-\alpha y, x-\alpha y) = ||x||^2 - 2\alpha||x||\,||y|| + ||y||^2\alpha^2 = (||x|| - \alpha||y||)^2.$$

Take $\alpha = ||x||/||y|| > 0$ to obtain $x = \alpha y$. ■

2. Examples of Hilbert spaces

7.1. Space $\mathbf{R}^n$ if for the elements

$$x = (x_1, x_2, \ldots, x_n), \qquad y = (y_1, y_2, \ldots, y_n)$$

an inner product is introduced by the formula

$$(x,y) = \sum_{i=1}^{n} x_i y_i. \tag{7.6}$$

7.2. Space l_2 with the inner product

$$(x,y) = \sum_{i=1}^{\infty} x_i y_i.$$

7.3. Spaces $L_2(a,b)$ and $L_2(\Omega)$ with corresponding inner products

$$(f,g) = \int_a^b f(x)g(x)dx \quad \text{and} \quad (f,g) = \int_\Omega f(x)g(x)d\Omega. \tag{7.7}$$

7.4. Let $w(x)$ be positive integrable on a segment $[a,b]$ function. Then it is possible to construct, in a similar way, along with the space

$L_2(a,b)$ a real space $L_2((a,b),w)$ where an inner product is defined by the formula

$$(f,g) = \int_a^b f(x)\,g(x)\,w(x)\,dx. \tag{7.8}$$

Remark. Inner products for complex Hilbert spaces in Examples 7.1–7.4 are to be taken respectively as

$$\sum_{i=1}^{n} x_i\bar{y}_i, \ \sum_{i=1}^{\infty} x_i\bar{y}_i, \ \int_a^b f(x)\overline{g(x)}dx, \ \int_\Omega f(x)\overline{g(x)}d\Omega, \ \int_a^b f(x)\overline{g(x)}w(x)dx.$$

3. Orthogonality and linear independence. Sonin–Schmidt orthogonalization process. Orthogonal and orthonormed systems. Examples

Two nonzero elements $x, y \in H$ are referred to as *orthogonal* ($x \perp y$) to each other if $(x,y) = 0$. If, besides, the elements x, y are *normed*; that is, $\|x\| = \|y\| = 1$, then they are referred to as *orthonormed.* Two sets H' and H'' from H are referred to as *orthogonal* ($H' \perp H''$) if $(h', h'') = 0$ for all $h' \in H'$, $h'' \in H''$. A set of elements $H' \subset H$ is referred to as *orthonormed* (*orthonormed system*) if its elements are normed and pairwise orthogonal.

Theorem 1. *If a sequence of elements $x_1, x_2, \ldots, x_n, \ldots$ ($x_i \neq 0$) from H forms an orthogonal system, they are linearly independent.*

□ Let us prove it by contradiction. Let there exist the numbers $n_1, n_2, \ldots, n_k$ and scalars $\lambda_1, \lambda_2, \ldots, \lambda_k$ ($\sum\limits_{i=1}^{k} |\lambda_i| > 0$) such that

$$\sum_{i=1}^{k} \lambda_i x_{n_i} = 0.$$

Take an inner product of this equality and x_{n_j}, $i \leq j \leq k$ to obtain $\lambda_j(x_{n_j}, x_{n_j}) = 0$; since $(x_{n_j}, x_{n_j}) > 0$, then $\lambda_j = 0$, $j = 1, 2, \ldots, k$. ■

Theorem 2. *Any system $x_1, x_2, \ldots, x_n, \ldots$ of linearly independent elements from H can be transformed into an orthogonal system*

$e_1, e_2, \dots, e_n, \dots$ *and orthonormed system* $f_1, f_2, \dots, f_n, \dots$ *with the help of the Sonin–Schmidt orthogonalization process.*

□ Assign $e_1 = x_1$ and let $e_2 = x_2 - \gamma_{21} e_1$. Select the number γ_{21} so that $e_2 \perp e_1$. Then by the equality $0 = (e_1, x_2) - \gamma_{21}(e_1, e_1)$ we obtain $\gamma_{21} = (e_1, x_2)/(e_1, e_1)$; in addition, $\|e_2\| \neq 0$ since otherwise the elements x_1 and x_2 would be linearly dependent. Let $e_1, e_2, \dots, e_{k-1}$ be already constructed. Take e_k in the form

$$e_k = x_k - \sum_{i=1}^{k-1} \gamma_{ki} e_i \tag{7.9}$$

and select the numbers γ_{ki} so that e_k is orthogonal to $e_1, e_2, \dots, e_{k-1}$. Take in turn an inner product of equality (7.9) and elements e_j, $j = 1, 2, \dots, k-1$ to obtain that

$$\gamma_{kj} = (e_j, x_k)/(e_j, e_j)$$

and so forth. We get an orthogonal system $e_1, e_2, \dots, e_n, \dots$ with this algorithm; an orthogonal system can be obtained by the formulas $f_n = e_n/\|e_n\|$ $n = 1, 2, \dots$. ■

4. Examples

7.4. A system of trigonometric functions 1, $\cos nt$, $\sin nt$, $n = 1, 2, \dots$ is orthogonal in the space $L_2(0, 2\pi)$.

7.5. If we orthogonalize an aggregate of powers $1, x, x^2, \dots, x^n, \dots$ in real space $L_2((a, b), w)$ with inner product (7.8), then we arrive at a system of polynomials $p_0(x) = \text{const}$, $p_1(x)$, $p_2(x)$, $\dots$, $p_n(x)$, $\dots$ *orthonormed with a weight* $w(x)$:

$$\int_a^b p_n(x)\, p_m(x)\, w(x)\, dx = \delta_{nm}. \tag{7.10}$$

For $a = -1$, $b = 1$ we obtain with accuracy up to multipliers the following polynomials: Legendre polynomials for $w(x) = 1$, Chebyshev polynomials for $w(x) = (1 - x^2)^{\mp 1/2}$ of the first and the second kind respectively; Chebyshev–Hermite polynomials for $a = -\infty$, $b = \infty$, $w(x) = e^{-x^2}$; Chebyshev–Laguerre polynomials for $a = 0$, $b = \infty$, $w(x) = e^{-x}$.

5. Symmetric and Hermitian bilinear forms. Formal construction of new (energetic) spaces

Let $D(M)$ be everywhere dense in a set H and let *symmetric bilinear form* $M(x,y)$ be assigned on the elements $x, y \in D(M)$ which means that any pair of elements $x, y \in D(M)$ has corresponding number $M(x,y)$ (real for real space H) and this correspondence possesses the following properties.

(a) $M(x,y) = M(y,x)$;

(b) $M(\lambda x, y) = \lambda M(x,y)$;

(c) $M(x_1 + x_2, y) = M(x_1, y) + M(x_2, y)$ for $\forall\ x, y, x_1, x_2 \in D(M)$ and arbitrary real λ.

The *quadratic form* of bilinear form $M(x,y)$ is a function $M(x,x)$ assigned on $D(M)$. Bilinear form $M(x,y)$ is referred to as *positive definite* if there exists the number $C > 0$ such that for $\forall\ x \in D(M)$ the following inequality is valid

$$M(x,x) \geq C\|x\|^2. \tag{7.11}$$

Let $D(M) = H$; then bilinear form $M(x,y)$ is referred to as *Hermitian* and if (7.11) is satisfied, it can be taken as the (new) inner product in H:

$$[x,y]_M = M(x,y). \tag{7.12}$$

The corresponding (new) norm is then determined by the equality

$$[x]_M = (M(x,x))^{1/2}. \tag{7.13}$$

A set forming H is a Hilbert set with a new inner product.

Example 7.7. Let $H = \mathbf{R}^n$ with inner product (7.6) and let A be a symmetric positive definite matrix. Assign $M(x,y) = (Ax,y)$; then $[x,y]_M = (Ax,y)$.

Now let now a bilinear form be positive definite and $D(M) \neq H$. Form a new Hilbert space H_M according to the following construction. Define the new inner product on $D(M)$ by formula (7.12) and a norm by formula (7.13) and then supplement $D(M)$ in a norm $[\cdot]_M$. As a result of this operation, we obtain a new Hilbert space H_M which is often referred to as *energetic*; its frame (everywhere dense part) consists of elements $D(M)$ of old Hilbert space H, which are supplemented with new limit elements.

6. Sobolev spaces $W_2^1(\Omega)$

Let $H = L_2(\Omega)$ with inner product (7.7) and $D(M) = C^1(\bar{\Omega})$ form a set everywhere dense in $L_2(\Omega)$. Take

$$M(u,v) = \int\limits_{\Omega} \left(\sum_{i=1}^{n} \frac{\partial u}{\partial x_i} \cdot \frac{\partial v}{\partial x_i} + uv \right) d\Omega \tag{7.14}$$

for $\forall\ u, v \in C^1(\Omega)$. Then H_M denoted in this case as $W_2^1(\Omega)$ will be one of the so-called *Sobolev spaces* which are described shortly in Section 14 of Chapter 2. Another notation is frequently used for this space: $H^1(\Omega)$.

§ 8. Problems on the Best Approximation. Orthogonal Expansions and Fourier Series in a Hilbert Space

Problem on the search for the best approximation by elements of convex set. Expansion into a sum of orthogonal subspaces. Fourier series, minimal property of Fourier coefficients. Bessel inequality, Parseval–Steklov equality. Complete orthogonal systems, orthogonal expansions, and separability. Isomorphism and isometry of Hilbert spaces.

1. Problem I on the search for the best approximation

Take Problem I stated in Section 3 on the best approximation of an assigned element $x \in H$ by elements of approximating set $M \subset H$. It is remarkable that a Hilbert space, being complete and possessing a distinctive metric and the concept of orthogonality, allows one to completely solve this problem for a rather wide class of sets M.

Let a set M be assigned in Hilbert space H and let a point $x \in H$; define a distance from the point x to the set M by formulas (1.20) and (6.1):

$$\rho(x, M) = \inf_{u \in M} \|x - u\|. \tag{8.1}$$

Theorem 1. *Let M be a closed convex set in H and let $x \notin M$. Then there exists a unique element $y \in M$ such that $\rho(x, M) = \|x-y\|$* (See Figure 6).

□ By Lemma 3 from Section 1, $d = \rho(x, M) > 0$ and by the definition $\rho(x, M)$ will be found $u_n \in M$ such that

$$d^2 \leq \|x - u_n\|^2 < d^2 + \frac{1}{n}. \tag{8.2}$$

The sequence u_n is fundamental. Indeed, making use of the parallelogram equality substituting there $x - u_n$ for x and $x - u_m$ for y obtains

$$\|u_n - u_m\|^2 + \|2x - u_n - u_m\|^2 = 2(\|x - u_n\|^2 + \|x - u_m\|^2).$$

On the other hand, taking into account that $(u_n + u_m)/2 \in M$ since M is convex and that $\|x - (u_n + u_m)/2\| \geq d$ we get

$$\|u_n - u_m\|^2 = 2(\|x - u_n\|^2 + \|x - u_m\|^2) - 4\left\|x - \frac{u_n + u_m}{2}\right\|^2$$

$$\leq 2\left(\left(d^2 + \frac{1}{n}\right) + \left(d^2 + \frac{1}{m}\right)\right) - 4d^2 = 2\left(\frac{1}{n} + \frac{1}{m}\right)$$

whence the fundamentality of u_n follows. By virtue of its closeness of M $\{u_n\}$ converges to some element $y \in M$. Passing to a limit in formula (8.2) we obtain $\|x - y\| = d$.

Let us prove the uniqueness of the element y. Let for some $\bar{y} \in M$ $\|x - \bar{y}\| = d$ too. Then by the parallelogram equality

$$4d^2 = 2(\|x - y\|^2 + \|x - \bar{y}\|^2) = \|y - \bar{y}\|^2 + 4\left\|x - \frac{y + \bar{y}}{2}\right\|^2$$

$$\geq \|y - \bar{y}\|^2 + 4d^2, \quad \text{that is,} \quad \|y - \bar{y}\| = 0 \quad \text{and} \quad y = \bar{y}. \ ■$$

The best approximation $y \in M$ is sometimes called a *projection on M of the element $x \notin M$*. Properties of an element of the best approximation are characterized by the following theorem (Figure 6).

Theorem 2. *A point y is a projection of a point x on closed convex set M iff for $\forall\ z \in M$ the following expression holds*

$$(z - y, x - y) \leq 0. \tag{8.3}$$

□ *Necessity.* Let y be a projection of x on M and let $z \in M$. Then for $\forall\ \lambda \in [0,1]$

$$v = (1-\lambda)y + \lambda z \in M$$

since M is convex and $\|x-y\|^2 \leq \|x-v\|^2$. Open this inequality

$$\|x-y\|^2 \leq \|(x-y)-\lambda(z-y)\|^2 = \|x-y\|^2+\lambda^2\|z-y\|^2-2\lambda(x-y,z-y);$$

that is, $\lambda^2\|z-y\|^2 - 2\lambda(x-y,z-y) \geq 0$ for $\forall\ \lambda \in [0,1]$ which is possible only under two conditions: $\lambda = 0$, but then $v = y$; and $\lambda > 0$, $(x-y, z-y) \leq 0$.

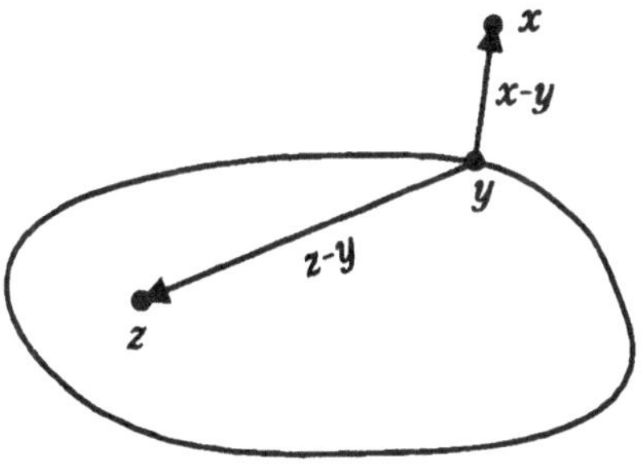

Figure 6

Sufficiency. Let inequality (8.3) be valid for some point $y \in M$ and let $\forall\, z \in M$. Then $\|z-x\|^2 = \|(z-y) + (y-x)\|^2 = \|z-y\|^2 + \|y-x\|^2 + 2(z-y, y-x) \geq \|y-x\|^2$ for $\forall\ z \in M$ which corresponds to the definition of a projection. ■

2. Expansion of a Hilbert space into a sum of orthogonal subspaces

Let $M = L$ be a subspace in H; that is, closed linear space, and let $x \in H$, but $x \notin L$. Since L is a closed convex set, then the corollaries following from the Theorems 1 and 2 are valid.

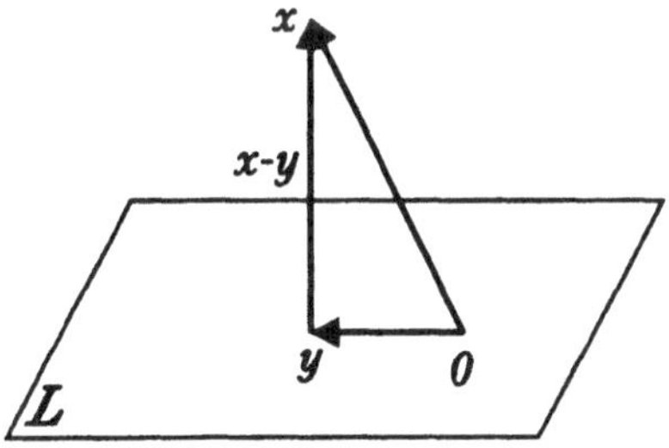

Figure 7

Corollary 1. *There exists a unique element $y \in L$* (Figure 7) *realizing a distance from a point x to the subset L:*

$$\rho(x, L) = \|x-y\|.$$

Corollary 2. *Let $\|x-y\| = \rho(x,L)$, then $x-y \perp L$.*

□ Assign in Equation (8.3) for $y \neq 0$: $z=0$ and then $z=2y$ and assign for $y=0$: $z=z_0 \neq 0$ and $z=-z_0$. ■

An important corollary is valid which we formulate as a theorem.

Theorem 3. *Let L be a subspace in H; then for $\forall\ x \in H$ the expansion*

$$x = y + z \tag{8.4}$$

is valid where $y \in L$, $z \perp L$. This expansion is unique.

□ Take y defined by x in Corollary 1 (if $x \in L$, then $y=0$) and assign $x = y + (x-y)$ where $z = x-y \perp L$ by Corollary 2. ■

The element y in expansion (8.4) is called *orthogonal projection* x on the subspace L.

Corollary 3. *The Pyphagor theorem is valid:*

$$\|x\|^2 = \|y\|^2 + \|z\|^2$$

□ since $(y,z)=0$. ■

Let L be a linear manifold in H (not necessarily closed). A set of all elements from H orthogonal to L is called the *orthogonal supplement* to L and is denoted by $L^\perp$.

Lemma 2. *$L^\perp$ is a subspace in H.*

□ Prove the linearity of $L^\perp$. Let $z_1, z_2 \in L^\perp$; that is, $(z_1,y) = (z_2,y) = 0$ for $\forall\, y \in L$. Then for any scalars λ_1, λ_2

$$(\lambda_1 z_1 + \lambda_2 z_2, y) = \lambda_1(z_1,y) + \lambda_2(z_2,y) = 0$$

for $\forall\ y \in L$; that is, $\lambda_1 z_1 + \lambda_2 z_2 \in L^\perp$.

Prove the closeness of $L^\perp$. Let $\{z_n\} \in L^\perp$ be assigned and let $z_n \to z \in H$, $n \to \infty$. We have for $\forall\ y \in L$: $(z_n,y)=0$. Pass to a limit here for $n \to \infty$ with regard to the continuity of the inner product to obtain $(z,y)=0$ for $\forall\ y \in L$; that is, $z \in L^\perp$. ■

Theorem 4. *If there is no nonzero element orthogonal to all elements of L, that is, if $L^\perp = \{0\}$, then it is necessary and sufficient for the manifold L to be everywhere dense in H.*

□ *Necessity.* Obviously, it follows from $x \perp L$ that $x \perp \bar{L}$ (recall, that the overbar here means the closure of a set). But, by definition $\bar{L} = H$ and consequently $x \perp H$, in particular, $x \perp x$, whence $x = 0$.

Sufficiency. Let L be not everywhere dense in H. Then $\bar{L} \neq H$ and there exists an element $x \notin \bar{L}$. By Theorem 3 we have $x = y + z$ where $y \in \bar{L}$, $z \perp \bar{L}$ and since $x \notin \bar{L}$, then $z \neq 0$ which contradicts the condition. ■

If L is a subspace in H, then $L^{\perp}$ is a subspace in H too. We say, keeping in mind single-valued expansion (8.4) for $\forall\, x \in H$, that space H is expandable into a direct sum of two orthogonal subspaces L and $L^{\perp}$ and write down this fact as

$$H = L \dot{+} L^{\perp}. \tag{8.4'}$$

3. Fourier series in Hilbert space. Minimal property of Fourier coefficients

Let $\{\varphi_k\}_1^\infty \in H$ be an orthogonal system $\varphi_k \neq 0$ and $(\varphi_k, \varphi_j) = 0$ for $k \neq j$. A series of type $\sum\limits_{k=1}^{\infty} \alpha_k \varphi_k$ is called a *series with respect to* $\{\varphi_k\}_1^\infty$. Let $x \in H$. The numbers $C_k = (x, \varphi_k)/\|\varphi_k\|^2$, $k = 1, 2, \ldots$ are called *Fourier coefficients* of an element x, the series

$$\sum_{k=1}^{\infty} C_k \varphi_k \tag{8.5}$$

is called the *Fourier series* for x with respect to the orthogonal system $\{\varphi_k\}_1^\infty$, and $\sum\limits_{k=1}^{n} C_k \varphi_k$ is called the *Fourier polynomial* of the element x.

Take again Problem I on the search for an element of the best approximation; take a subspace L_n as an approximating set which is spanned on the first n elements of the orthogonal system: $\varphi_1, \varphi_2, \ldots, \varphi_n$. Let $u_n = \sum\limits_{k=1}^{n} \alpha_k \varphi_k$ and $x \notin L_n$. Calculate the value

$$d_n^2 = \|x - u_n\|^2.$$

We get, accounting for the orthogonality:

$$d_n^2 = (x, x) - 2\sum_{k=1}^{n} \alpha_k (x, \varphi_k) + \sum_{k=1}^{n} \alpha_k^2 (\varphi_k, \varphi_k).$$

But $(x, \varphi_k) = C_k \|\varphi_k\|^2$, consequently,

$$d_n^2 = \|x\|^2 + \sum_{k=1}^{n} \alpha_k^2 \|\varphi_k\|^2 - 2\sum_{k=1}^{n} \alpha_k C_k \|\varphi_k\|^2$$

$$= \|x\|^2 - \sum_{k=1}^{n} C_k^2 \|\varphi_k\|^2 + \sum_{k=1}^{n} (\alpha_k - C_k)^2 \|\varphi_k\|^2. \quad (8.6)$$

Compute now

$$\rho_n = \rho(x, L_n) = \inf_{u_n \in L_n} \|x - u_n\| = \inf_{\alpha_1, \ldots, \alpha_n} d_n.$$

Explicit form (8.6) for d_n shows that $\inf d_n$ is attained for $\alpha_k = C_k$, $k = \overline{1, n}$. This property of Fourier coefficients $C_1, C_2, \ldots, C_k$ is called the *minimal property of Fourier coefficients.* So, the following theorem is valid.

Theorem 5. *Let $\{\varphi_n\}_1^\infty$ be orthogonal in the H system and let L_n be a subspace in H spanned on $\varphi_1, \ldots, \varphi_n$. Then $\rho_n = \rho(x, L_n)$, $x \in H$ is assigned by the following formulas*

$$\rho_n = \left\| x - \sum_{k=1}^{n} C_k \varphi_k \right\|, \quad \rho_n^2 = \|x\|^2 - \sum_{k=1}^{n} C_k^2 \|\varphi_k\|^2, \quad (8.7)$$

where C_k, $k = \overline{1, n}$, are Fourier coefficients of the element x with respect to the system $\{\varphi_n\}_1^\infty$.

Corollary 4. If $m > n$, then

$$\left\| x - \sum_{k=1}^{m} C_k \varphi_k \right\| \leq \left\| x - \sum_{k=1}^{n} C_k \varphi_k \right\|.$$

So, we have established an important fact: the best approximation of an element $x \in H$ by elements of L_n is a Fourier polynomial of the element x: $\sum_{k=1}^{n} C_k \varphi_k$.

Corollary 5. *To solve Problem* IV *on the approximation of assigned set $C \subset H$ with nth Fourier polynomials means finding the value*

$$G(C, L_n, H) = \sup_{x \in C} \left[\|x\|^2 - \sum_{k=1}^{n} \frac{(x, \varphi_k)^2}{\|\varphi_k\|^2} \right].$$

4. Bessel inequality, Parseval–Steklov equality. Complete orthogonal systems

Since $\rho_n^2 \geq 0$, we get by formula (8.7)

$$\sum_{k=1}^{n} C_k^2 \|\varphi_k\|^2 \leq \|x\|^2.$$

Pass to a limit in this inequality for $n \to \infty$ to obtain the inequality

$$\sum_{k=1}^{\infty} C_k^2 \|\varphi_k\|^2 \leq \|x\|^2 \tag{8.8}$$

which is called the *Bessel inequality.*

Let us now discuss the convergence of a Fourier series. An orthogonal system $\{\varphi_k\}_1^\infty$ from Hilbert space H is referred to as *complete* if for $\forall\ x \in H$ $\sum\limits_{k=1}^{\infty} C_k \varphi_k = x$ (the Fourier series for x converges to x).

The complete orthogonal system $\{\varphi_k\}_1^\infty$ is termed an *orthogonal basis* in H. By inequalities (8.7) and (8.8) we get

Corollary 6. *It is necessary and sufficient for $\{\varphi_k\}_1^\infty$ to be complete, that for $\forall\ x \in H$*

$$\sum_{k=1}^{\infty} C_k^2 \|\varphi_k\|^2 = \|x\|^2. \tag{8.9}$$

This equality is called the *Parseval–Steklov equality.* The following theorem is valid.

Theorem 6. *The orthogonal system $\{\varphi_k\}_1^\infty$ from H is complete, iff its linear hull L is everywhere complete in H* (i.e., $\bar{L} = H$).

□ Let $\{\varphi_k\}_1^\infty$ be complete. If $\bar{L} \neq H$, then $\exists\ x_0 \in \bar{L}^\perp$, $x_0 \neq 0$. But then $C_k = (x_0, \varphi_k)/\|\varphi_k\|^2 = 0$ and by virtue of completeness

$x_0 = \sum\limits_{k=1}^{\infty} C_k\varphi_k = 0$. The contradiction thus obtained shows that $\bar{L} = H$.

Now let $\bar{L} = H$ and $x \in H$. Then for $\forall \varepsilon > 0\ \exists x_\varepsilon \in L$ such that $\|x - x_\varepsilon\| < \varepsilon$. Since $x_\varepsilon \in L$ then, consequently, x_ε is a linear combination with respect to $\{\varphi_k\}_1^\infty$; that is, $x_\varepsilon = \sum\limits_{k=1}^{N} \alpha_k\varphi_k$. Therefore (see Theorem 5 and Corollary 4), for $n > N$

$$\left\| x - \sum_{k=1}^{n} C_k\varphi_k \right\| \leq \left\| x - \sum_{k=1}^{N} C_k\varphi_k \right\| \leq \|x - x_\varepsilon\| < \varepsilon$$

which means that $\sum\limits_{k=1}^{\infty} C_k\varphi_k = x$ for $\forall\ x \in H$. ■

5. Orthogonal expansions and separability. Isomorphism and isometry of separable Hilbert spaces

Theorem 7. *Existence of an orthogonal basis of a finite or countable number of elements is necessary and sufficient for Hilbert space H to be separable.*

□ *Necessity.* Since H is separable, countable, and everywhere dense, set $\{\bar{x}_n\}$ will be found in it. Let $\bar{x}_k$ be the first nonzero element in $\{\bar{x}_n\}$; assign $x_1 = \bar{x}_k$. Let $\bar{x}_l$ be the first element in the sequence $\bar{x}_{k+1}, \bar{x}_{k+2}, \ldots$ linearly independent of $\bar{x}_1$; then assign $x_2 = \bar{x}_l$. Take the sequence $\bar{x}_{l+1}, \bar{x}_{l+2}, \ldots$ and let x_3 be its first element that is not a linear combination of x_1 and x_2. Proceeding in this way we obtain a finite or infinite system of elements $\{x_n\}$ whose linear hull L contains the system $\{\bar{x}_n\}$ which is everywhere dense in H. Orthogonalizing $\{x_n\}$ we arrive at orthogonal complete $\{e_n\}$ since linear hulls of systems $\{x_n\}$ and $\{e_n\}$ coincide. Thus, the system $\{e_n\}$ forms a basis in H.

Sufficiency. A set of all finite linear combinations of elements of a basis $\{e_n\}$ with rational coefficients forms a countable everywhere dense in H set. ■

Theorem 8. *Any real separable Hilbert space H is isometric and isomorphic to real space l_2 and, consequently,* all (*real*) *separable Hilbert spaces are isometric and isomorphic to each other.*

□ Let H be a separable Hilbert space and let $\{e_n\}$ be a complete orthonormed system in H. Then it is possible for all $x \in H$ to find a sequence of numbers C_k that are Fourier coefficients of an element x with respect to the system $\{e_n\}$ so that $\sum_{k=1}^{\infty} C_k^2 = \|x\| < \infty$; consequently, $\tilde{x} = (C_1, C_2, \ldots, C_n, \ldots) \in l_2$. Each element $x \in H$ has thus some element $\tilde{x} \in l_2$ and, by virtue of the completeness condition for $\{e_n\}$, $\|x\|_H = \|\tilde{x}\|_{l_2}$. It is obvious that if $\tilde{y} \in l_2$ corresponds to $y \in H$, then for any real λ and μ, an element $\lambda x + \mu y \in H$ will have corresponding element $\lambda\tilde{x} + \mu\tilde{y} \in l_2$ and

$$\|\lambda x + \mu y\|_H = \|\lambda\tilde{x} + \mu\tilde{y}\|_{l_2}. \tag{8.10}$$

Now let $\tilde{z} = (\xi_1, \xi_2, \ldots, \xi_n, \ldots)$ be an arbitrary element of l_2. Take in H the elements $z_n = \sum_{k=1}^{n} \xi_k e_k$, $n = 0, 1, \ldots$. For $n > m$ we have

$$\|z_n - z_m\|^2 = \sum_{k=m+1}^{n} \xi_k^2 \to 0 \quad \text{for} \quad n, m \to \infty.$$

Thus, $\{z_n\}$ is fundamental in H and converges to some $z \in H$. Since $(z, e_k) = \lim_{n\to\infty} (z_n, e_k) = \xi_k$, then

$$z = \sum_{k=1}^{\infty} \xi_k e_k \tag{8.11}$$

which means that each element $\tilde{z} \in l_2$ corresponds to some $z \in H$ according to formula (8.11). So, we have one-to-one correspondence between the elements of spaces H and l_2 preserving operations of addition and multiplication by a scalar; that is, the spaces H and l_2 are isomorphic. Formula (8.10) shows that this correspondence is the isometry too. ■

Corollary 7 (the Riesz–Fisher theorem). Real spaces $L_2(0,1)$ and l_2 are isometric and isomorphic (Figure 8).

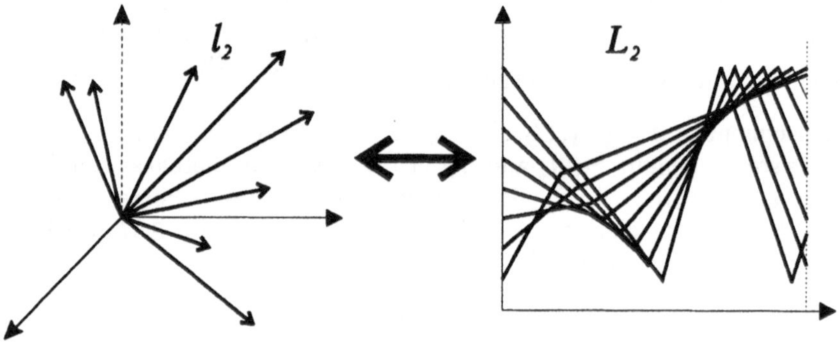

Figure 8

Remark. This is a good occasion, with the agreement in mind that isomorphic and isometric spaces are indistinguishable, to recall Poincare's remark that mathematics is a skill of calling different things by the same name.

§ 9. Some Extremal Problems in Normed and Hilbert Spaces

Problem I in normed space, existence and uniqueness. Systems of equations with the Gram matrix in Problem I in Hilbert space. Properties of Gram determinants, positiveness, generalized Cauchy–Bunyakovskii inequality. Estimation of a deviation. Problem I in space $C[a,b]$ if a set A consists of polynomials. The least deviation. Theorem on existence and uniqueness of best approximation by polynomials and alternance theorem.

In many divisions of mathematics problems appear on the best (in one sense or another) approximation of an assigned function by functions of a certain class. Some such problems in abstract statements were listed in Section 3.

1. Problem I in normed space. Aspects of existence and uniqueness of an element of the best approximation

Let E be normed space (not necessarily complete) and let L be its finite-dimensional closed subspace. Take an element $x \in E$ such that

$x \notin L$. Then, as is known, for each $y \in L$ the value $\|y - x\|$ is called a *deviation* and the element $y \in L$ for which the deviation $\|y - x\|$ has the least value, is called an *element of the best approximation* (in the subspace L).

Theorem 1. *For $\forall\, x \in E$ in finite-dimensional (closed) subspace $L \subset E$ there exists an element of the best approximation.*

□ It is sufficient to consider the case when $x \notin L$ since otherwise the element x itself would be an element of the best approximation. Take $\forall\, y_0 \in L$ and let $\sigma = \|x - y_0\|$. It is clear that an element of the best approximation is to be searched for among the elements $y \in L$ that belong to closed ball $\overline{D}_\sigma(x)$. Let $M = \overline{D}_\sigma(x) \cap L$; the set M is a closed bounded set lying in finite-dimensional space; therefore, it is bicompact.

We see, regarding the deviation as a functional $f(y) = \|x - y\|$ of y, that $f(y)$ is a continuous functional. By Theorem 3 from Section 2 there exists $y^* \in M$ such that $f(y)$ has the least value for $y = y^*$. This very element y^* is the element of the best approximation for x. ■

The space L can be often generated by n linearly independent elements $x_1, x_2, \ldots, x_n$; that is, represening their linear hull. Then Problem I is reduced to the search for coefficients $\lambda_1, \lambda_2, \ldots, \lambda_n$, such that

$$\rho = \rho(x, L) = \min_{\lambda_1, \lambda_2, \ldots, \lambda_n} \left\| x - \sum_{k=1}^{n} \lambda_k x_k \right\| \tag{9.1}$$

is attained; that is, the linear combination of elements $x_1, x_2, \ldots, x_n$, is to be found, whose deviation from assigned elements x is the least one. Theorem 1 guarantees the existence of such a combination.

Uniqueness of an element of the best approximation can be proved under some additional conditions. The following lemma is valid.

Lemma 1. *An element $y^* \in L$ of the best approximation is determined uniquely in strictly normed space E.*

□ Let along with y^* there exist one more element of the best approximation for x: $\bar{y} \in L$ and $f(y^*) = f(\bar{y}) = \bar{\sigma}$. Then the element $(1/2)(y^* + \bar{y})$ will be such an element also, since

$$\left\| x - \frac{1}{2}(y^* + \bar{y}) \right\| = \frac{1}{2}\|x - y^*\| + \frac{1}{2}\|x - \bar{y}\| = \bar{\sigma} \tag{9.2}$$

[an inequality is impossible in (9.2) since the left-hand side cannot be smaller than the right-hand side]. By (9.2) there exists $\lambda > 0$ such that

$$x - y^* = \lambda(x - \bar{y}).$$

For $x \notin L$ the last equality is possible just for $\lambda = 1$, and then $y^* = \bar{y}$. ■

2. Problem (9.1) in a Hilbert space. Properties of Gram determinants, generalized Cauchy–Bunyakovskii inequality. Estimation of deviation

Let $E = H$ be a Hilbert space. The problem of the search for an element of the best approximation $y^* \in L$ is reduced in this case to the solution of a certain system of linear algebraic equations of nth order. In fact, let

$$y^* = \sum_{k=1}^{n} \lambda_k x_k; \tag{9.3}$$

then by Corollary 2 from Section 8

$$(y^* - x, x_i) = 0, \qquad i = \overline{1, n} \tag{9.4}$$

since $y^* - x \perp L$, and $x_i \in L$. Substituting Equation (9.3) into (9.4) we obtain, to determine λ_k, a system of linear algebraic equations of the following type

$$\sum_{k=1}^{n} \lambda_k (x_k, x_i) = (x, x_i), \qquad i = \overline{1, n}. \tag{9.5}$$

The determinant of this system

$$|(x_k, x_i)| = G(x_1, x_2, \ldots, x_n) \tag{9.6}$$

is called the *Gram determinant.*

Since the space H is strictly normed and $\{x_k\}_1^n$ is a linearly independent system, the system for any x by Theorem 1 and by Lemma 1 has one and just one solution; that is,

$$G(x_1, x_2, \ldots, x_n) \neq 0.$$

Let us find an expression for a square of deviation ρ:

$$\rho^2 = (x - y^*, x - y^*) = (x - y^*, x) - (x - y^*, y^*) = (x - y^*, x) = (x, x) - (y^*, x)$$

or
$$\sum_{k=1}^{n} \lambda_k (x_k, x) = (x, x) - \rho^2. \tag{9.7}$$

Join this equation to system (9.5); since the supplemented system of $n+1$ equations with n unknowns has a solution, we conclude that Equation (9.7) is linearly dependent with n equations (9.5); that is, the following determinant is equal to zero.

$$\begin{vmatrix} (x_1, x_1) & \vdots & (x_n, x_1), & (x, x_1) - 0 \\ (x_1, x_2) & \vdots & (x_n, x_2), & (x, x_2) - 0 \\ \dots & \dots & \dots & \dots \\ (x_1, x_n) & \vdots & (x_n, x_n), & (x, x_n) - 0 \\ (x_1, x) & \vdots & (x_n, x), & (x, x) - \rho^2 \end{vmatrix} = 0.$$

Represent the last column as a difference of two vectors to obtain

$$\rho^2 = \frac{G(x_1, x_2, \dots, x_n, x)}{G(x_1, x_2, \dots, x_n)} > 0. \tag{9.8}$$

By this relationship and by the fact that $G(x_1) = (x_1, x_1) > 0$ $(x_k \neq 0)$ for $n = 1$ the Gram determinant is always greater than or equal to zero; it is equal to zero iff the elements $\{x_k\}_1^n$ in (9.6) are linearly dependent. So, for $\forall\ x_i \in H$

$$G(x_1, x_2, \dots, x_n) \geq 0. \tag{9.9}$$

This inequality is the generalization of the Cauchy–Bunyakovskii inequality, since for $n = 2$ we have

$$G(x_1, x_2,) = (x_1, x_1)(x_2, x_2) - (x_1, x_2)^2 \geq 0.$$

3. Problem I for the space $C([a, b], w(x))$. The least deviation. Theorems on best approximation and on alternance

The problem on the best uniform approximation of an arbitrary continuous function by polynomials was studied for the first time by famous Russian mathematician P. Chebyshev in the middle of the 19th century. His works on polynomials of the best approximation served as a basis for the development of a new mathematical discipline: the constructive theory of functions.

The space $C([a,b],w)$, where $w(x)>0$ and is continuous on an $[a,b]$ function, is not strictly normed. In order to ensure the uniqueness of the solution to Problem I it is imperative that the system of functions $x_1(x), x_2(x), \ldots, x_n(x)$ in (9.1) satisfy additional conditions – so-called *Chebyshev conditions*: any generalized polynomial $\sum_{k=1}^{n} \alpha_k x_k(x)$ with $\sum_{k=1}^{n} \alpha_k^2 > 0$ is to have not more than $n-1$ zeros on $[a,b]$. The system of $n+1$ functions $x_k(x) = x^{k-1}$, $k = 1, 2, \ldots, n+1$ satisfies these conditions; therefore, the following theorem is valid.

Theorem 2. *Let a function $f(x)$ be continuous on a segment $[a,b]$. Then for any n there exists a unique polynomial*

$$P_n(x) = \sum_{i=0}^{n} a_i x^i$$

that among all polynomials $Q_n(x)$ of nth power is the least deviating with the weight $w(x)$ from the function $f(x)$:

$$P_n(x) = \arg \inf_{Q_n(x)} \max_{x \in [a,b]} |(f(x) - Q_n(x))w(x)|. \tag{9.10}$$

The existence of such a polynomial follows from Theorem 1; we do not present a proof of uniqueness here. A value of the least deviation, denoted for the space $C([a,b],w)$ as E_n, is determined by the formula

$$E_n = E_n(f) = \inf_{Q_n(x)} \max_{x \in [a,b]} |(f(x) - Q_n(x))w(x)|. \tag{9.11}$$

It is obvious that $E_n \le E_{n+1}$.

The *first Weierstrass theorem* says: if $f(x) \in C[a,b]$, then it is possible for $\forall \varepsilon > 0$ to find a polynomial $P_n(x)$ of power $n = n(\varepsilon)$ for which the inequality

$$\max_{x \in [a,b]} |(f(x) - P_n(x))w(x)| < \varepsilon$$

is valid. By this theorem $E_n \to 0$ for $n \to \infty$.

Let $\eta(x) = (f(x) - P_n(x))w(x)$ where $P_n(x)$ is a polynomial of nth power. The following theorem is valid.

Theorem 3 (the main Chebyshev theorem on alternance). *It is necessary and sufficient for a polynomial $P_n(x)$ to be the least deviating from a function $f(x) \in C([a,b],w(x))$ in a metric of the space*

$C([a,b], w(x))$, that there exist $n+2$ points $\xi_1 < \xi_2 < \cdots < \xi_{n+2} \in [a,b]$ where $\max\limits_{x\in[a,b]} |\eta(x)|$ is attained with sequential alternation of sign.

□ Let us prove the sufficiency. Let there exist a polynomial of nth power $Q_n(x)$ that is strictly the least deviating from $f(x)$ as compared to $P_n(x)$. Then polynomial of nth power $v_n(x) = Q_n(x) - P_n(x) = f(x) - P_n(x) - (f(x) - Q_n(x))$ will have in the alternance points the same sign as the function $\eta(x)$. Since the latter changes the sign $n+1$ times inside $[a,b]$, then $v_n(x)$ is to have $n+1$ zeros; that is, $Q_n(x) = P_n(x)$. The contradiction proves the sufficiency of the theorem's conditions. ■

We have left an error in Figure 9 illustrating the alternance theorem for $w(x) \equiv 1$ and $n = 5$. Where is the error?

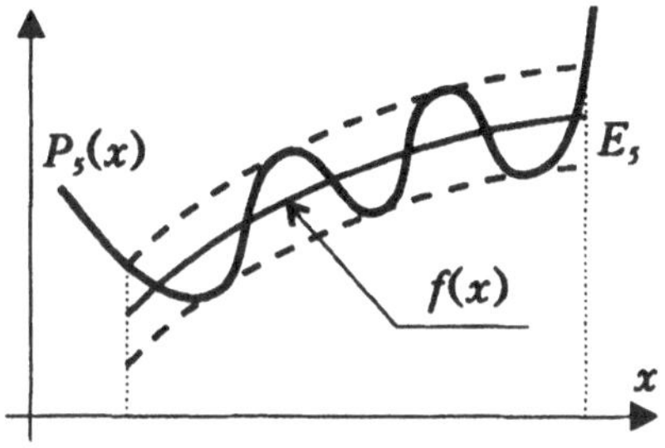

Figure 9

Unfortunately, if normed spaces are not Hilbert spaces, there is no simple algorithm to find an element of the best approximation for an arbitrary element of these spaces. In particular, such are the spaces $C([a,b], w(x))$.

§ 10. Polynomials the Least Deviating from Zero. Chebyshev Polynomials and Their Properties

Chebyshev polynomials and their properties. Solution of a Problem I type: on approximation of Ax^n by polynomials of power $n-1$; on construction of polynomials, the least deviating from zero, with fixed coefficient at highest power or fixed value in some point.

The polynomials discovered by Chebyshev and called by Bertrand the "miracle of analysis" lie in the basis of the solution of many extremal problems that appear in various divisions of mathematics and its applications: theory of best approximation, integration, optimization of iteration methods, optimal control, and others.

Let Π_n be a class of polynomials of nth power with fixed coefficient $A \neq 0$ at x^n. Take the following problem: find

$$P_n(x) = \arg \inf_{Q_n(x)\in\Pi_n} \max_{x\in[-1,1]} |Q_n(x)w(x)|, \qquad (10.0)$$

where $w(x) > 0$ in (–1,1) is continuous on a [–1,1] weight function. Let us call the polynomials $P_n(x)$ (10.0) polynomials *the least deviating on* [–1,1] *from zero with the weight* $w(x)$ or, in short, *extremal* polynomials. Obviously, they are defined with accuracy up to the multiplier A. Since each polynomial $Q_n(x) \subset \Pi_n$ has a form $Q_n(x) = Ax^n - q_{n-1}(x)$, where $q_{n-1}(x)$ is a polynomial of power $n-1$, then it follows by the alternance theorem that it is sufficient for $P_n(x)$ to be extremal polynomial and that the function $P_n(x)w(x)$ would have on [–1,1] $n+1$ alternance point. Consequently, the extremal polynomial $P_n(x)$ has n different roots lying in (–1,1).

1. Chebyshev polynomials and their properties

Chebyshev polynomials of the first kind denoted as $T_n(x)$ are defined with the formulas:

$$T_0(x) = 1, \qquad T_1(x) = x \qquad (10.1)$$

$$T_{n+1}(x) = 2xT_n(x) - T_{n-1}(x), \qquad n = 1, 2, \dots . \qquad (10.2)$$

It follows by (10.2) that

$$T_2(x) = 2x^2 - 1, \qquad T_3(x) = 4x^3 - 3x, \dots \qquad (10.3)$$

and $T_n(x)$ is a polynomial of nth power with a coefficient at x^n equal to 2^{n-1} for $n \geq 1$.

Chebyshev polynomials can be represented in several explicit forms. Here is the most remarkable of them, one that determines many properties of these polynomials.

Theorem 1. For real x

$$T_n(x) = \cos(n \arccos x) \qquad (10.4)$$

if $|x| \leq 1$ and

$$T_n(x) = \operatorname{ch}(n \operatorname{arcch} x) \qquad (10.5)$$

if $|x| > 1$.

□ Let us prove the consistency of formulas (10.4), (10.5) and (10.1), (10.2), respectively. In fact, for $|x| \leq 1$,

$$\cos(0 \cdot \arccos x) = \cos 0 = 1 = T_0(x)$$

$$\cos(1 \cdot \arccos x) = x = T_1(x).$$

The validity of recursive formula (10.3) follows from trigonometric identity $\cos(n+1)\theta + \cos(n-1)\theta = 2\cos\theta \cos n\theta$ if

$$\theta = \arccos x \tag{10.6}$$

is substituted there.

Formula (10.5) is verified similarly with a symbol ch substituted for cos. ■

Remark 1. It is possible to give (for those familiar with the theory of a function of complex variables), a single formula for $T_n(x)$, namely, formula (10.4) where x may be arbitrary, complex one included; to this end corresponding branches of functions in formula (10.4) must be selected. Write down this formula once more:

$$T_n(x) = \cos(n\theta), \qquad x = \cos\theta, \qquad 0 \leq Re\theta \leq \pi. \tag{10.7}$$

It is easy to see, that polynomials $T_n(x)$ are even functions for even n and odd functions for odd n and $\max\limits_{-1 \leq x \leq 1} |T_n(x)| = 1$.

Theorem 2. *For all (complex) x*

$$T_n(x) = \frac{1}{2}\left(\left(x + \sqrt{x^2-1}\right)^n + \left(x - \sqrt{x^2-1}\right)^n\right)$$

$$= \frac{1}{2}\left(\left(\sqrt{\frac{x+1}{2}} + \sqrt{\frac{x-1}{2}}\right)^m + \left(\sqrt{\frac{x+1}{2}} - \sqrt{\frac{x-1}{2}}\right)^m\right), \tag{10.8}$$

where a branch is chosen for the function $\sqrt{z}$ such that $\sqrt{1} = 1$ and $m = 2n$.

□ Expressions containing radicals in formulas (10.8) are annihilated when their terms are raised to the nth power. Check again, to prove the theorem, the consistency of formulas (10.8) and formulas (10.1) and (10.2). ■

Rewrite (10.8) using the equality $(x-\sqrt{x^2-1})^n = (x+\sqrt{x^2-1})^{-n}$ as

$$T_n(x) = \frac{1}{2}((x+\sqrt{x^2-1})^n + (x+\sqrt{x^2-1})^{-n}). \tag{10.9}$$

Chebyshev polynomials satisfy many identities. We present two of them.

Lemma 1. *The following identities are valid for m, n:*

$$T_m(x)T_n(x) = \frac{1}{2}(T_{m+n}(x) + T_{m-n}(x)), \tag{10.10}$$

where $m \geq n$ and

$$T_m(T_n(x)) = T_{mn}(x). \tag{10.11}$$

□ Formula (10.10) is a result of summing of right- and left-hand sides of two trigonometric identities:

$$\cos m\theta \cos n\theta \pm \sin m\theta \sin n\theta = \cos(m \mp n)\theta$$

with regard to relationship (10.6). Formula (10.11) follows from (10.7):

$$T_m(T_n(x)) = \cos(m \arccos(\cos(n \arccos x)))$$

$$= \cos(mn \arccos x) = T_{mn}(x). \ ■$$

It is remarkable that zeros of a Chebyshev polynomial and its maxima on the segment $[-1,1]$ are determined with explicit formulas. Denote

$$t_{nk} = x_k = \cos\theta_k, \qquad \theta_k = \frac{2k-1}{2n}\pi, \qquad k = \overline{1,n}. \tag{10.12}$$

Lemma 2. *For each integer $n > 0$ zeros of polynomial $T_n(x)$ are real, simple, enclosed in the interval* (–1,1), *and are expressed with formulas* (10.12).

□ Substitute numbers θ_k from (10.12) into formula (10.7). ■

Similarly, it follows from (10.7) that in the points

$$\bar{x}_i = \cos\frac{i\pi}{n}, \qquad i = \overline{0, n} \tag{10.13}$$

$|T_n(\bar{x}_i)| = 1$ or, more exactly,

$$T_n(\bar{x}_i) = (-1)^i, \qquad i = \overline{0, n}. \tag{10.14}$$

Figure 10 shows diagrams on $[0, 1]$ for the first Chebyshev polynomials.

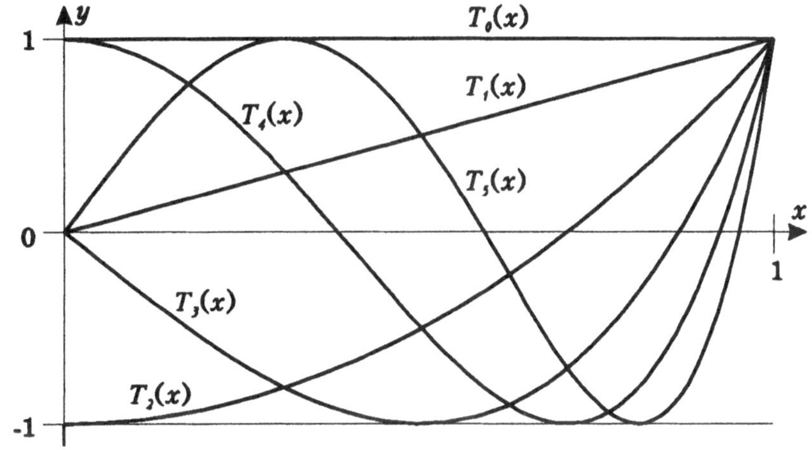

Figure 10

Theorem 3. *A sequence of Chebyshev polynomials of the first kind $T_n(x)$ is orthogonal on the segment $[-1, 1]$ with the weight $(1-x^2)^{-1/2}$. The corresponding orthonormed sequence consists of the polynomials*

$$\frac{1}{\sqrt{\pi}}T_0(x), \qquad \sqrt{\frac{2}{\pi}}\,T_k(x), \qquad k = 1, 2, \ldots .$$

□ Calculate the integrals

$$I_{kl} = \int_{-1}^{1} T_k(x)\,T_l(x)(1-x^2)^{-1/2}dx.$$

Use formula (10.4) and substitute $x = \cos\theta$ to obtain

$$I_{kl} = \int_0^\pi \cos k\theta \cos l\theta d\theta$$

$$= \frac{1}{2}\int_0^\pi (\cos(k-l)\theta + \cos(k+l)\theta)\, d\theta = \begin{cases} \pi, & k = l = 0 \\ \pi/2, & k = l \neq 0 \\ 0, & k \neq l \end{cases} . \blacksquare$$

We get the following corollary, taking into account extremal properties of the Fourier series.

Corollary 1. *For arbitrary integer $N > 0$ and arbitrary function $f(x) \in L_2((-1,1),w)$, where $w = (1-x^2)^{-1/2}$ polynomial of the best approximation of nth power in the space $L_2((-1,1),w)$ equals*

$$\sum_{k=1}^{n} a_k(f)T_k(x), \tag{10.15}$$

where

$$a_k(f) = \gamma_k \int_{-1}^{1} f(x)T_k(x)(1-x^2)^{-1/2}dx, \qquad \gamma_0 = \frac{1}{\pi},$$

$$\gamma_k = \frac{2}{\pi}, \qquad k = 1, 2, \ldots . \tag{10.16}$$

Formulas (10.15) and (10.16) for sufficiently smooth functions from $C[-1,1]$ prove to present approximations close to the best ones in $C[-1,1]$.

The second linearly independent with respect to variable n solution of recursive relationships

$$u_{n+1}(x) = 2xu_n(x) - u_{n-1}(x), \qquad n = 1, 2, \ldots \tag{10.17}$$

is represented by Chebyshev polynomials of the second kind $U_n(x)$ of power n that have the trigonometric form:

$$U_n(x) = \frac{\sin(n+1)\theta}{\sin\theta}$$

$$= \frac{1}{2}\left((x+\sqrt{x^2-1})^{n+1} - (x-\sqrt{x^2-1})^{n+1}\right)/\sqrt{x^2-1} \tag{10.18}$$

$$U_1(x) = 2\cos\theta, \qquad U_0(x) = 1, \qquad U_{-1}(x) = 0.$$

The numbers
$$u_{nk} = \cos\frac{i\pi}{n+1}, \qquad i=\overline{1,n} \tag{10.19}$$
will be roots of $U_n(x)$. Differentiate (10.7) as a complicated function to obtain
$$T_n'(x) = nU_{n-1}(x). \tag{10.20}$$

By formulas (10.3), (10.7), (10.11), and (10.18) we get that
$$T_{3n}(x) = T_3(T_n(x)) = T_n(x)(2T_n(T_2(x)) - 1) \tag{10.21}$$
and
$$U_{nm-1}(x) = U_{m-1}(T_n(x))U_{n-1}(x) \tag{10.22}$$
which show that the roots of $U_{nm-1}(x)$ consist of roots of the polynomials $U_{m-1}(T_n(x))$ and $U_{n-1}(x)$, and one third of roots $T_{3n}(x)$ are roots of the polynomial $T_n(x)$.

Represent in three forms the general solution of equations (10.16) for each x:
for $n \geq 0$
$$u_n(x) = C_1(x)T_n(x) + C_2(x)U_{n-1}(x) \tag{10.23}$$
for $n \geq l-1$, where $l > 0$ is integer
$$u_n(x) = C_1(x)T_{n-l+1}(x) + C_2(x)U_{n-l}(x), \tag{10.24}$$
where $C_1(x), C_2(x)$ are own for each formula arbitrary functions, and for $-1 \leq x = \cos\theta \leq 1$, $n \geq 0$
$$u_n(x) = \frac{1}{w(x)}\cos(n\theta + \Psi(\theta)) \tag{10.25}$$
in trigonometric form, where $w(x) \neq 0$ in (-1,1) and $\Psi(\theta)$ is an arbitrary bounded function.

There is a large series of extremal problems in the space $C[-1,1]$ whose solutions are expressed through Chebyshev polynomials.

2. Solution of Problem I: (a) on the best approximation of Ax^n by a polynomial of (n − 1)th power; (b), (c) on construction of polynomials the least deviating from zero with fixed coefficients at highest power or with fixed value in some point

Problem I (a). Find among polynomials of power $n-1$ a polynomial of type

$$q_{n-1}(x) = \sum_{k=0}^{n-1} a_k x^k \tag{10.26}$$

the least deviating in $C[-1,1]$ from the function $f(x) = Ax^n$, where $A \neq 0$, that is, find a solution to the problem

$$q_{n-1}^0(x) = \arg\min_{q_{n-1}} \max_{|x|\leq 1} |Ax^n - q_{n-1}(x)|. \tag{10.27}$$

Denote

$$P_n(x) = Ax^n - \sum_{k=0}^{n-1} a_k x^k. \tag{10.28}$$

Let us show that a solution to this problem [explicit expression for the polynomial $q_{n-1}^0(x)$] follows from the statement: the polynomial $P_n(x)$ in (10.28) is to be such that

$$P_n(x) = \frac{A}{2^{n-1}} T_n(x). \tag{10.29}$$

□ In fact, in this case a difference between an approximated function Ax^n and a polynomial of power $n-1$ that approximates it satisfies the alternance theorem conditions. Consequently,

$$q_{n-1}^0(x) = A\left(x^n - \frac{1}{2^{n-1}} T_n(x)\right). \blacksquare \tag{10.30}$$

This reasoning is a base for the statement that expression (10.29) is also a solution of the following problem.

Problem I (b). Find among all polynomials of nth power with a coefficient $A \neq 0$ at x^n a polynomial the least deviating from zero on [-1,1].

One may, knowing the explicit solution of Problem I (a) suggest a method referred to as telescopic: the convolution of partial sums of a power series. Let a function $f(x)$ be expanded for $-1 \leq x \leq 1$ into the Taylor series

$$f(x) = \sum_{j=0}^{\infty} a_j x^j$$

and let n be known such that an error of the formula

$$f(x) \approx P_n(x) = \sum_{j=0}^{n} a_j x^j, \tag{10.31}$$

where $a_n \neq 0$ is sufficiently small. Substitute the term $a_n x^n$ in (10.31) with a polynomial of power $n-1$ by formula (10.30) for $A = a_n$ and with an error equal to $a_n/2^{n-1}$ obtain a new approximation to $f(x)$ by a polynomial of power $n-1$

$$f(x) \approx P_{n-1}(x) = \sum_{j=0}^{n-1} a_j x^j + a_n \left(x^n - \frac{1}{2^{n-1}} T_n(x) \right). \tag{10.32}$$

If an error of new approximation is small enough, we repeat the process of decreasing the power and so forth.

Problem I (c) is presented as the theorem proved by Chebyshev in 1881 for a segment $[a, b]$.

Theorem 4. *Among all polynomials $Q_n(x) = a_0 x^n + a_1 x^{n-1} + \cdots + a_n$ with real coefficients, that satisfy the condition $Q_n(\xi) = R > 0$ ($\xi \notin [a, b]$ is real), the polynomial*

$$P_n(x) = RT_n\left((2x - a - b)/(b - a)\right)/T_n(\Theta) \tag{10.33}$$

and just this polynomial has the least deviation from zero equal to $|R/T_n(\Theta)|$ where $\Theta = (2\xi - a - b)/(b - a)$.

□ We may assume, with no loss of generality, that $a = -1$ and $b = +1$, since by substituting a variable by the formula

$$t = (2x - a - b)/(b - a) \tag{10.34}$$

we can always reduce the general case to the proposed one. Assume for the same reason that $\Theta = \xi > 1$. Then

$$P_n(x) = \frac{R}{T_n(\Theta)} T_n(x) \tag{10.35}$$

and

$$E_n = \max_{-1 \leq x \leq 1} |P_n(x)| = \frac{R}{|T_n(\Theta)|}. \tag{10.36}$$

Let us show that a solution to Problem I (c) is assigned by formula (10.35). Indeed, if another polynomial $\bar{P}_n(x)$ is a solution to this problem and $\bar{E}_n = \max_{-1 \leq x \leq 1} |\bar{P}_n(x)|$ is such that $\bar{E}_n \leq E_n$, then forming the difference

$$\varphi_n(x) = P_n(x) - \bar{P}_n(x),$$

we obtain that the polynomial $\varphi_n(x)$ of nth power turns to zero in $n+1$ points.

In fact, $\varphi_n(\Theta) = 0$. If $\bar{E}_n < E_n$, then $\varphi_n(x)$ changes the sign n more times between the points $\bar{x}_k$ [see (10.13)] which are maximum points of $|P_n(x)|$. Consequently, $\varphi_n(x) \equiv 0$. If $\bar{E}_n = E_n$, then either $\varphi(\bar{x}_{k_i}) = 0$, $i = 1, 2, \ldots, m$ $(m < n)$ and then $\varphi'(\bar{x}_{k_i}) = 0$; that is, each point $\bar{x}_{k_i}$ is a repeated zero of the polynomial $\varphi_n(x)$, or, in the opposite case, $\operatorname{sign} \varphi_n(\bar{x}_k) = \operatorname{sign} P_n(\bar{x}_k)$. A thorough count of the number of changes of signs of the function $\varphi_n(x)$ and of the number of its zeros will show again that $\varphi_n(x) \equiv 0$. ∎

The following Corollary follows from Theorem 4.

Corollary 1. *Among all polynomials $Q_n(x)$ of nth power, satisfying the condition*

$$\max_{x \in [a,b]} |Q_n(x)| \leq \eta \quad (\eta > 0) \tag{10.37}$$

the polynomial

$$\tilde{P}_n(x) = \eta T_n\left(\frac{2x-a-b}{b-a}\right) \tag{10.38}$$

has the largest absolute value in the point $\xi \notin [a,b]$ with accuracy up to a sign.

□ The corollary is easily proved by contradiction. ∎

Let $\xi > b$. Add and subtract a value η from $\tilde{P}_n(x)$ to arrive at the following corollary.

Corollary 2. *The polynomials*

$$P_n^{\pm}(x) = \frac{1}{2}(\tilde{P}_n(x) \pm \eta) \tag{10.39}$$

assume maximal value in the point $\xi > b$ among all polynomials of nth power satisfying respectively on $[a,b]$ the conditions:

$$\textit{for} \quad P_n^{+}(x) \quad 0 \leq P_n^{+}(x) \leq \eta \tag{10.40}$$

$$\textit{for} \quad P_n^{-}(x) \quad -\eta \leq P_n^{-}(x) \leq 0. \tag{10.41}$$

Corollary 1 can be supplemented by the following theorem, which we present here skipping the proof.

Theorem 5. *For any polynomial* $Q_n(x)$ *of nth power satisfying condition* (10.37) *the inequality*

$$\left|\frac{dQ_n}{dx}\right|_{x=\xi} \leq \eta \left|\frac{dT_n(2x-a-b)/(b-a)}{dx}\right|_{x=\xi} \tag{10.42}$$

is valid in any point $\xi \notin [a,b]$.

§ 11. Some Extremal Polynomials

Polynomial solutions of recursive relationships, ChMBS-polynomials, asymptotic formulas, iteration method for finding the roots of extremal polynomials, complex case. Solution of Problem V on the best approximation by interpolation Lagrange polynomial.

1. Polynomial solutions of recursive relationships. ChMBS-polynomials

Let us go back to formulas (10.23)–(10.25) that represent general solutions of equations (10.17). *Assume* that the functions $C_1(x)$, $C_2(x)$, $w(x)$, and $\psi(\theta)$ are selected continuous and such that the solution $u_n(x)$ *is a polynomial of nth power.* Then by (10.25)

$$u_n(x)w(x) = \cos(n\theta + \psi(\theta)). \tag{11.1}$$

If $\psi(\pi) - \psi(0) \geq 0$ in (11.1), then the product $u_n(x)w(x)$ has on [-1,1] at least $n+1$ alternance points, assumes there the values ± 1 alternatively, and consequently all roots of the polynomial $u_n(x)$ are different and lie in [-1,1]. It means that the polynomial $u_n(x)$ is a solution to problem (9.10) for $A = A_n$, where A_n is a coefficient of $u_n(x)$ for x^n; that is, the polynomial deviating least from zero with a weight $w(x)$ on the segment [-1,1]. Its roots x_i satisfy the equations

$$x_i = \cos\theta_i, \quad \theta_i = \frac{1}{n}\left(\frac{2i-1}{2}\pi - \psi(\theta_i)\right), \quad i = 1, \ldots, n\,. \tag{11.2}$$

The roots, if the function $\psi(\theta)$ satisfies the Lipschitz conditions with a constant $L < n$, can be found with the method of successive approximations (4.3):

$$\theta_i^{j+1} = \frac{1}{n}\left(\frac{2i-1}{2}\pi - \psi(\theta_i^j)\right), \qquad \theta_i^0 = \frac{2i-1}{2n}\pi, \tag{11.3}$$

where $i = 1, \ldots, n$, $j = 0, 1, \ldots$.

Chebyshev investigated the following remarkable cases of extremal polynomials.

(1) For $\psi(\theta) \equiv 0, \quad w(x) \equiv 1 : u_n(x) = T_n(x)$.

(2) For $\psi(\theta) = \theta - \pi/2, \quad w(x) = \sqrt{1-x^2} : u_n(x) = U_n(x)$.

(3) For $\psi(\theta) = \theta/2, \quad w(x) = \sqrt{1+x}$:

$$V_n(x) = \frac{\cos(n+1/2)\theta}{\cos\frac{\theta}{2}}. \tag{11.4}$$

(4) For $\psi(\theta) = (\theta - \pi)/2, \quad w(x) = \sqrt{1-x}$:

$$W_n(x) = \frac{\sin(n+1/2)\theta}{\sin\frac{\theta}{2}}. \tag{11.5}$$

We show that formulas (11.4) and (11.5) define polynomials of nth power; they are called the Chebyshev polynomials of the third and the fourth kind, respectively. Obviously, the roots of all the preceding Chebyshev polynomials are found explicitly with formulas (11.2).

We make use of some elementary facts from the theory of a complex variable, for example, that $\exp(i\alpha) = \cos\alpha + i\sin\alpha$; as a rule, we take that branch of a two-valued function $\sqrt{z}$, for which $\sqrt{1} = 1$, otherwise the conditions determining a branch will be assigned in $\{\cdot\}$. Let $0 \le Re\,\theta \le \pi$ $\{\frac{\pi}{2} = \arccos 0\}$ in (10.6) and let

$$h(x) = \sqrt{\frac{x+1}{2}} + \sqrt{\frac{x-1}{2}} = \exp\frac{i\theta}{2}; \tag{11.6}$$

then it is easily verified that

$$h^{-1}(x) = \sqrt{\frac{x+1}{2}} - \sqrt{\frac{x-1}{2}} = \exp\frac{-i\theta}{2}. \tag{11.7}$$

Consider the two following functions for an integer or semiinteger ν

$$\mathrm{Co}_\nu(x) = \frac{1}{2}(h^{2\nu}(x) + h^{-2\nu}(x)) \tag{11.8}$$

$$\mathrm{Si}_\nu(x) = \frac{1}{2}(h^{2\nu}(x) - h^{-2\nu}(x)). \tag{11.9}$$

Let us call them *Chebyshev functions of νth power of the first and the second order*, respectively. By (11.6) and (11.7)

$$\mathrm{Co}_\nu(x) = \mathrm{Cos}\,\nu\theta, \qquad \mathrm{Si}_\nu(x) = i\,\mathrm{Sin}\,\nu\theta. \tag{11.10}$$

The functions $\mathrm{Co}_\nu(x)$ and $\mathrm{Si}_\nu(x)$ are homogeneous 2νth power functions of the kind

$$\sum_{j+k=\nu} Aij(\sqrt{x+1})^j(\sqrt{x-1})^k. \tag{11.11}$$

Hence, if ν is integer, then only terms with even j, k are present in (11.11) for $\mathrm{Co}_\nu(x)$, since the rest of terms annihilate; that is, $\mathrm{Co}_\nu(x)$ is a polynomial of power ν of x; for the same reason only terms with odd j, k are present in (11.11) for $\mathrm{Si}_\nu(x)$; that is, $\mathrm{Si}_\nu(x)$ is a polynomial of $(\nu-1)$th power multiplied by $\sqrt{x^2-1}$. If ν is semiinteger, then only terms with j odd and k even are present in (11.11) for $\mathrm{Co}_\nu(x)$; that is, $\mathrm{Co}_\nu(x)$ is a polynomial of $(\nu-1/2)$th power multiplied by $\sqrt{x+1}$, and only terms with even j and odd k are present in (11.11) for $\mathrm{Si}_\nu(x)$; that is, $\mathrm{Si}_\nu(x)$ is a polynomial of $(\nu-1/2)$th power multiplied by $\sqrt{x-1}$.

Therefore, the following functions are polynomials.

$$T_n(x) = \mathrm{Co}_n(x) \tag{11.12}$$

$$U_n(x) = \mathrm{Si}_n(x)/\sqrt{x^2-1} \tag{11.13}$$

$$V_n(x) = \mathrm{Co}_{n+1/2}(x)/\sqrt{(x+1)/2} \tag{11.14}$$

$$W_n(x) = \mathrm{Si}_{n+1/2}(x)/\sqrt{(x-1)/2}. \tag{11.15}$$

Remark. The alternance theorem prompts one more way to obtain separate extremal polynomials containing power functions in weight functions. Let a polynomial $P_n(x)$ be extremal polynomial with a weight $w(x)$ on the segment [-1,1]. Select $m+1 < n+1$ points of alternance: $\xi_k, \xi_{k+1}, \ldots, \xi_{k+m}$, where $k \geq 1$, $k+m \leq n+1$. Denote by y_i, $i = 1, 2, \ldots, m$ the zeros of $P_n(x)$ lying in $[\xi_k, \xi_{k+m}]$. The rest of the zeros of $P_n(x)$ are denoted by z_i $i = 1, 2, \ldots, n-m$. Let

$$p_m(x) = \prod_{i=1}^{m}(x-y_i), \qquad q_{n-m}(x) = \prod_{i=1}^{n-m}(x-z_i)$$

$$d = \min\{1-\xi_{k+m}, \min_{z_i>\xi_{k+m}}(z_i-\xi_{k+m})\}, \quad c = \min\{\xi_k+1, \min_{z_i<\xi_k}(\xi_k-z_i)\}.$$

Then according to the alternance theorem for $\xi_k - c \leq a \leq \xi_k$, $\xi_{k+m} \leq b \leq \xi_{k+m} + d$ the polynomial $p_m(x)$ will be an extremal polynomial for $A = 1$ on the segment $[a, b]$ for the weight function $\bar{w}(x) = w(x)|q_{n-m}(x)|$. It is possible to map the segment $[a, b]$ onto segment $[-1, 1]$ with transformation (10.34). For example, the $(n-2)$th power polynomial

$$T_n(x \cos \frac{\pi}{2n})/(x^2 - 1) \tag{11.16}$$

will be extremal on $[-1, 1]$ for $\bar{w}(x) = |1 - x^2|$, and the $(n-1)$th power polynomials

$$T_n\left(x \cos^2 \frac{\pi}{4n} \pm \sin^2 \frac{\pi}{4n}\right)/(x \pm 1) \tag{11.17}$$

will be extremal on [-1,1] for $w(x) = |1 \pm x|$, respectively.

2. ChMBS-polynomials

Formula (10.24) is a source of one more series of polynomials. Let $S_{l-1}(x)$, $S_l(x)$ be polynomials of powers $l-1$, l, respectively. Define in (10.24) the functions $C_1(x)$, C_2 so that

$$u_{l-1}(x) = S_{l-1}(x), \qquad u_l(x) = S_l(x). \tag{11.18}$$

Take $n = l-1, l$ in (10.24) taking into account that $T_1(x) = x$, $T_0(x) = 1$, $U_0(x) = 1$, and $U_{-1}(x) = 0$ to obtain for the determination of $C_1(x), C_2$ the system of equations

$$C_1(x) = S_{l-1}(x), \qquad C_1(x)x + C_2(x) = S_l(x);$$

that is,

$$C_1(x) = S_{l-1}(x), \quad C_2(x) = S_l(x) - xS_{l-1}(x). \tag{11.19}$$

Then for $n \geq l-1$ the function

$$u_n(x) = S_{l-1}(x)T_{n-l+1}(x) + (S_l(x) - xS_{l-1}(x))U_{n-l}(x) \tag{11.20}$$

is a polynomial of power n. Let us reduce this polynomial to trigonometric form (10.25). If we denote $g(x) = (xS_{l-1}(x) - S_l(x))/(1-x^2)$

$$w(x) = \sqrt{S_{l-1}^2(x) + g^2(x)} \tag{11.21}$$

and assign a phase function $\psi(\theta)$ with the inequalities

$$\cos((l-1)\theta + \psi(\theta)) = \frac{S_{l-1}(\cos\theta)}{w(\cos\theta)}$$

$$\sin((l-1)\theta + \psi(\theta)) = \frac{g(\cos\theta)}{w(\cos\theta)},$$

then formula (11.20) takes for $n \geq l-1$ the form (10.25) where the function $w(x)$ is determined by formula (11.21) and the function $\psi(\theta)$ equals

$$\psi(\theta) = \operatorname{Arctg}\frac{\cos\theta S_{l-1}(\cos\theta) - S_l(\cos\theta)}{\sin\theta S_{l-1}(\theta)} - (l-1)\,\theta. \qquad (11.22)$$

The continuous branch of Arctg is chosen here, for which Arctg $0 = 0$. Let us call these polynomials *ChMBS-polynomials* after mathematicians Chebyshev, Markov, Bernstein, and Szegö. One can show that under additional constraints for $S_{l-1}(x)$, $S_l(x)$ these polynomials have complete alternance; that is, are extremal polynomials. Coefficients of polynomials $S_{l-1}(x)$, $S_l(x)$ may be regarded as parameters; varying them, we obtain a series of extremal problems.

Let us obtain another important class of ChMBS-polynomials. Let $(l_1, l_2, \dots, l_m)$ be a set of m positive integer numbers and let $(a_1, a_2, \dots,$
$a_m)$ be a set of complex numbers such that $a_j^2 \notin [1,\infty) \cup \{0\}$. If a_j is complex, then this set contains the number $a_k = \bar{a}_j$ too (complex conjugate to a_j) with $l_k = l_j$. Let $l = \sum_{k=1}^{m} l_k$. Define a weight function with the formula

$$w(x) = \prod_{k=1}^{m} (1 - a_k T_{l_k}(x))^{-1/2} \qquad (11.23)$$

and find for this function an explicit form of the function $\psi(\theta)$ in the representation (10.25) and, consequently, the corresponding extremal ChMBS-polynomial. It is obvious that if a_i is real, then the expression in the kth brackets (11.23) is positive, and if $a_k = \bar{a}_j$, then a product of the expressions in brackets with indices k and j is positive also.

Define the function $\psi(\theta)$ as a sum:

$$\psi(\theta) = \sum_{k=1}^{m} \psi_k(\theta), \qquad (11.24)$$

where we define each continuous function $\psi_k(\theta)$ with the relationships:

$$\psi_k(0) = 0, \qquad k = 1, 1, \ldots, m, \tag{11.25}$$

$$\exp i\left(\frac{l_k\theta}{2} + \psi_k(\theta)\right) = \frac{1}{\sqrt{1 - a_k \cos l_k\theta}}\left(\sqrt{1 - a_k}\cos\frac{l_k\theta}{2}\right.$$
$$\left. + i\sqrt{1 + a_k}\sin\frac{l_k\theta}{2}\right), \qquad k = 1, 2, \ldots, m. \tag{11.26}$$

By (11.26)

$$\operatorname{tg}\left(\frac{l_k\theta}{2} + \psi_k(\theta)\right) = \sqrt{\frac{1 + a_k}{1 - a_k}}\operatorname{tg}\frac{l_k\theta}{2}, \qquad k = 1, 2, \ldots, m. \tag{11.27}$$

Find $\psi_k(\theta)$ from these equations; calculations show that for real a_k

$$\psi_k(\theta) = \operatorname{arctg}\frac{\sin l_k\theta}{A_k - \cos l_k\theta}, \tag{11.28}$$

where

$$A_k = a_k^{-1} + \sqrt{a_k^{-2} - 1}, \qquad \{|A_k| \geq 1\} \tag{11.29}$$

and if A_k and A_j are complex conjugate, then

$\psi_j(\theta) + \psi_k(\theta)$

$$= \operatorname{arctg}\frac{\sin l_k\theta((A_k + \bar{A}_k) - 2\cos l_k\theta)}{2\cos^2 l_k\theta - (A_k + \bar{A}_k)\cos l_k\theta + |A_k|^2 - 1}. \tag{11.30}$$

By (11.24), (11.28), and (11.30)

$$\psi(0) = \psi(\pi) = 0. \tag{11.31}$$

Let $\mu = n - l/2 \geq 0$ be an integer or semiinteger number; then use a formula for cosine and (11.26) to obtain

$$\cos(n\theta + \psi(\theta)) = \cos\left(\mu\theta + \tfrac{l\theta}{2} + \psi(\theta)\right)$$
$$= \frac{1}{2}\left(\exp(i\mu\theta)\prod_{k=1}^{m}\exp i\left(\frac{l_k}{2}\theta + \psi_k(\theta)\right)\right.$$
$$\left. + \exp(-i\mu\theta)\prod_{k=1}^{m}\exp\left(-i\left(\frac{l_k\theta}{2} + \psi_k(\theta)\right)\right)\right). \tag{11.32}$$

But, by (11.6) and (11.7)

$$\exp(i\mu\theta) = h^{2\mu}(x), \qquad \exp(-i\mu\theta) = h^{-2\mu}(x)$$

and by (11.26) and (11.10)

$$\exp \pm i\left(\frac{l_k\theta}{2} + \psi_k(\theta)\right) = \frac{\sqrt{1-a_k}\mathrm{Co}_{l_k/2}(x) \pm \sqrt{1+a_k}\mathrm{Si}_{l_k/2}(x)}{\sqrt{1-a_k T_{l_k}(x)}}.$$

So, we obtain with regard to (11.23) the formula (10.25)

$$C_n(x) = \frac{1}{2}\left[h^{2\mu}(x)\prod_{k=1}^{m}(\sqrt{1-a_k}Co_{l_k/2}(x) + \sqrt{1+a_k}Si_{l_k/2}(x))\right.$$

$$\left. + h^{-2\mu}(x)\prod_{k=1}^{m}(\sqrt{1-a_k}Co_{l_k/2}(x) - \sqrt{1+a_k}Si_{l_k/2}(x))\right] \qquad (11.33)$$

$$= \frac{1}{w(x)}\cos(n\theta + \psi(\theta)).$$

We can check with regard to formulas (11.6) and (11.7) for $h(x)$ and $h^{-1}(x)$, (11.8) and (11.9) for $Co_\nu(x)$ and $Si_\nu(x)$ that $C_n(x)$ is a homogeneous function of power $2n$ of kind (11.11) where only even values of j and k are present; that is, $C_n(x)$ is a polynomial of power n. It is a *ChMBS-polynomial of the first kind*; the coefficient A for x^n here is easily computed from the limit relationship

$$A = \lim_{x\to+\infty} C_n(x)/x^n. \qquad (11.34)$$

One more series of extremal polynomials is presented by Zolotaryev polynomials of the first and second types, of the first and second kinds that are expressed through elliptic functions.

3. Asymptotic formulas. Iteration method for finding the roots of extremal with a weight polynomials

It is possible to show in the general case that for a function $w(x)$ sufficiently smooth and positive on [-1,1] there exists a continuous phase function $\psi_n(\theta)$ of a definite class such that the polynomial

$P_n(x) \in \Pi_n$ deviating least from zero with a weight $w(x)$ on the segment [-1,1] is representable in the form

$$P_n(x) = \frac{E_n}{w(x)} \cos(n\theta + \psi_n(\theta)), \tag{11.35}$$

where E_n is a deviation from zero depending on the norming $P_n(x)$. Bernstein and Szegö obtained an asymptotic formula for the phase function $\psi_n(\theta) : \psi_n(\theta) \to \psi(\theta)$ for $n \to \infty$ where

$$\psi(\theta, w) = \psi(\theta) = \frac{\sin\theta}{\pi} \int_0^\pi \frac{\ln(w(\cos\theta)/w(\cos\varphi))}{\cos\theta - \cos\varphi} d\varphi \tag{11.36}$$

and the integral is understood in the sense of the main value. It is possible to verify that for Chebyshev polynomials of 1–4 kinds and for extremal ChMBS-polynomials, formulas (11.35) and (11.36) are exact for $\psi_n(\theta) = \psi(\theta)$; that is, they coincide with formula (10.25).

Representation (11.35), the asymptotic formula for the phase function (11.36), and formulas (11.2) and (11.3) for the roots are the basis for an iteration method for finding extremal polynomials that works effectively for large values of n.

Important properties of the phase function $\psi(\theta, w)$ follow from (11.36):

$$\psi(\theta, w_1 \cdot w_2) = \psi(\theta, w_1) + \psi(\theta, w_2) \tag{11.37}$$

$$\psi(\theta, w^\lambda) = \lambda\psi(\theta, w). \tag{11.38}$$

Using these properties and computing explicitly the integral in (11.36) we obtain that for

$$w(x) = \prod_{k=1}^{m} (1 - a_k T_{l_k}(x))^{\lambda_k} \tag{11.39}$$

$$\psi(\theta) = -2\sum_{k=1}^{m} \lambda_k \operatorname{arctg} \frac{\sin l_k\theta}{A_k - \cos l_k\theta}, \tag{11.40}$$

where A_k are determined in (11.29), and for

$$w(x) = \exp(c_0 + \sum_{k=1}^{m} c_k T_k(x)) \tag{11.41}$$

$$\psi(\theta) = \sum_{k=1}^{m} c_k \sin k\theta. \tag{11.42}$$

Let

$$P_n(x) = A\prod_{k=1}^{n}(x - x_k), \qquad (11.43)$$

where

$$x_0 = 1 > x_1 > x_2 > \cdots > x_n > -1 = x_{n+1},\ \theta_k = \arccos x_k. \quad (11.44)$$

We say that the polynomial $P_n(x)$ corresponds to the phase function $\psi_n(\theta)$ if its roots x_k are linked with relationship (11.44) and (11.2) for $\psi(\theta) = \psi_n(\theta)$ with roots θ_k of the function $\cos(n\theta + \psi_n(\theta))$. Calculations demonstrate that even the polynomial $P_n(x)$ corresponding to the phase function (11.36) yields a good approximation for extremal approximation to be found; we nevertheless, using property (11.37), extend this asymptotic formula to the computation of corrections to obtain approximate formulas for extremal polynomials in the following way.

Let a polynomial $Q_n(x)$ (an approximation to extremal one with a weight w) have the form (11.43) and correspond to a sufficiently smooth phase function $\psi_n(\theta)$. Then there exists a sufficiently smooth function $r_n(x) > 0$ for $x \in [-1, 1]$ (it is determined in a nonunique way) such that the polynomial $Q_n(x)$ is an exact extremal polynomial for a weight function $w_1(x) = w(x)\, r_n(x)$. Now employ the reasoning scheme applied in the inverse analysis and based on the asymptotic property (11.37) for *small corrections* to phase functions. Let us assume that $r_n(x)$ is a correction to the weight function $w(x)$ refining in asymptotic approximation the asymptotic formula (11.36). *Therefore, we call a polynomial* $\tilde{Q}_n(x)$ *of type* (11.43) *an asymptotic refinement of* $Q_n(x)$ *if the* $\tilde{Q}_n(x)$ *corresponds to a phase function* $\tilde{\psi}_n(\theta) = \psi_n(\theta) - \Phi_n(\theta)$, *where* $\Phi_n(\theta)$ *is* $\psi(\theta, r_n)$*-transformation* (11.36) *from the function* $r_n(x)$.

The formulated principle lies in the basis of the following iteration method for finding the roots of an extremal polynomial. Let us present this method schematically. Let $P_n^k(x)$ be an approximation obtained on the kth iteration step that has the form

$$P_n^k(x) = A\prod_{i=1}^{n}(x - x_i^k) \qquad (11.45)$$

and corresponds to the phase function $\psi(\theta) + \Phi_n^k(\theta)$, where $\psi(\theta) = \psi(\theta, w)$ is the function (11.36) and

$$x_i^k = \cos\theta_i^k, \quad \theta_i^k = \frac{1}{n}\left(\frac{2i-1}{2}\pi - \psi(\theta_i^k) - \Phi_n^k(\theta_i^k)\right), \qquad (11.46)$$

and

$$x_0^k = 1 > x_1^k > x_2^k > \cdots > x_n^k > -1 = x_{n+1}^k.$$

Let

$$\xi_i^k = \arg \max_{x \in [x_i^k, x_{i+1}^k]} |P_n^k(x) w(x)| \tag{11.47}$$

$$\lambda_i^k = |P_n^k(\xi_i^k) w(\xi_i^k)|, \qquad i = 0, 1, \ldots, n. \tag{11.48}$$

It is obvious that the values $P_n^k(\xi_i^k)w(\xi_i^k)$ have different signs in the neighboring points ξ_i^k. Select a class of correction weight functions $r_n^k(x)$ in the form

$$r_n^k(x) = \exp(G_n^k(x)), \tag{11.49}$$

where $G_n^k(x)$ is a polynomial of power n. Then according to (11.36)

$$\Phi_n^k(\theta) = \Psi(\theta, r_n^k).$$

In this case the integral in (11.36) is easily computed by formulas (11.42) if the function $r_n^k(x)$ is transformed in advance to the form (11.41) for $m = n$.

Let us make one more assumption: in order to equalize the maximums $|P_n^k(x)w(x)r_n^{k+1}(x)|$ on the segment [-1,1] with the help of the correction weight function $r_n^{k+1}(x)$ of type (11.49) it is sufficient, in the framework of asymptotic approximation, to equalize the values $|P_n^k(\xi_i^k)w(\xi_i^k)r_n^{k+1}(\xi_i^k)|$. Then the transition from $\Phi_n^k(\theta)$ to $\Phi_n^{k+1}(\theta)$ [and consequently from $P_n^k(x)$ to $P_n^{k+1}(x)$] in the proposed iteration method is performed with the following operations, having $P_n^k(x)$.

(1) determine ξ_i^k, λ_i^k by formulas (11.47) and (11.48) and compute $\ln \lambda_i^k$, $i = 0, 1, \ldots, n$;

(2) Compute the Lagrange polynomial $g_n^{k+1}(x)$ coinciding with the values $\ln \lambda_i^k$ in the points ξ_i^k and expand it into a sum with respect to Chebyshev polynomials of the first kind [see (11.41)]. Then the values of $|P_n^k(\xi_i^k)w(\xi_i^k)\exp(-g_n^{k+1}(\xi_i^k))|$ will be the same;

(3) compute the correction to the phase function

$$G_n^{k+1}(\theta) = \psi(\theta, \exp g_n^{k+1}(x)) \tag{11.50}$$

with formula (11.42);

(4) assume

$$\Phi_n^{k+1}(\theta) = \Phi_n^k(\theta) + \gamma_{k+1} G_n^{k+1}(\theta), \tag{11.51}$$

where $0 < \gamma_{k+1} \leq 1$ is an iteration parameter ($\gamma_{k+1} = 1$ for an exact asymptotic approximation);

(5) use the criterion $\max_i \lambda_i^k - \min_i \lambda_i^k \gtrless \varepsilon$, where ε is the accuracy, in order to decide, whether to continue the iterations or stop.

We have described the external iteration cycle; the internal cycle consists of finding the values θ_i^k for each k that are solutions to Equation (11.46) according to the scheme (11.3), where $\psi(\theta)+\Phi_n^k(\theta)$ is substituted for $\psi(\theta)$. The computation practice has demonstrated fast convergence of external and internal iterations for large n. Figure 11 shows the diagrams for $n = 20$ and $w(x) = 1/4 \exp x(1 - 0.5T_3(x))^2 \cdot(1 - 0.9T_2(x))^{-1/2}$: $P_{20}(x)w(x)$ with a solid line; $w(x)$ with a round dotted line; and $0.5\psi_{20}(\arccos x)$ with a small dotted line. The dispersion of maximums of (11.48) did not exceed 10^{-13} already at the fourth external iteration.

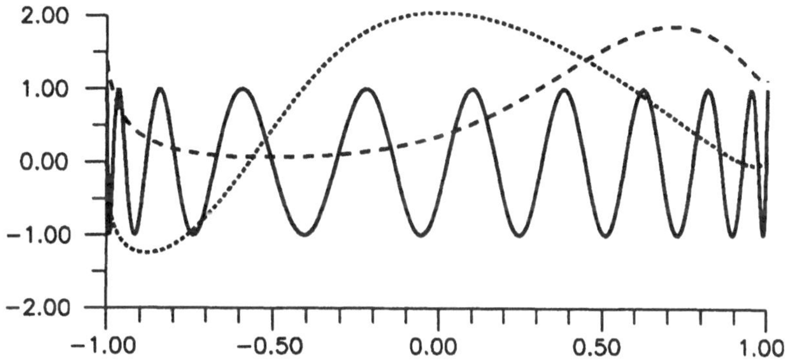

Figure 11

5. A class of extremal polynomials in a complex plane

Let us find extremal polynomials for some specific class of domains lying in the plane of a complex variable. Their parameters can be used while constructing iteration methods for problems with a complex spectrum lying in domains different from those traditionally used: segment or ellipse. To this end we again use the method of inverse analysis.

Let $z = x + iy$, $Q_m(z)$ be a polynomial of power m, $\rho > 0$, and let $\partial\Omega_\rho$ be a level line (lemniscate) for $Q_m(z)$:

$$\partial\Omega_\rho = \{z : |Q_m(z)| = \rho\} \tag{11.52}$$

limiting closed domain

$$\Omega_\rho = \{z : |Q_m(z)| \leq \rho\}. \tag{11.53}$$

The domain Ω_ρ is multiply connected for $m > 1$ and sufficiently small ρ. Let $z_0 \notin \Omega_\rho$ and $\overline{\prod}_N(z_0)$ be a class of polynomials $p_N(z)$ of power N satisfying the condition

$$p_N(z_0) = 1. \tag{11.54}$$

The following theorem is valid.

Theorem 1. For $\forall\, l > 0$

$$P_{ml}(z) = \arg \inf_{p_{ml} \in \overline{\prod}_{ml}(z_0)} \max_{z \in \Omega_\rho} |p_{ml}(z)| = \left(\frac{Q_m(z)}{Q_m(z_0)}\right)^l \tag{11.55}$$

and

$$\bar{E}_{ml}(z) = \inf_{p_{ml}(z) \in \overline{\prod}_N(z_0)} \max_{z \in \Omega_\rho} |p_{ml}(z)| = \left|\frac{\rho}{Q_m(z_0)}\right|^l. \tag{11.56}$$

□ A function $\varphi(z) = \tilde{P}_{ml}(z)/P_{ml}(z)$ is regular outside the domain Ω_ρ, $\varphi(z_0) = 1$ for any polynomial $\tilde{P}_{ml}(z) \in \overline{\prod}_{ml}(z_0)$ and different from $P_{ml}(z)$, and by virtue of the maximum module principle

$$\sup_{z \in \partial\Omega_\rho} |\varphi(z)| = \left|\frac{Q_m(z_0)}{\rho}\right|^l \sup_{z \in \partial\Omega_\rho} |\tilde{P}_{ml}(z)| > 1;$$

that is,

$$\sup_{z \in \partial\Omega_\rho} |\tilde{P}_{ml}(z)| > \left|\frac{\rho}{Q_m(z_0)}\right|^l. \ \blacksquare$$

The polynomials $P_{ml}(z)$ in (11.55) will be Faber polynomials for the domains Ω_ρ. Theorem 1 can be applied to an interesting class of polynomials $Q_m(z)$ whose roots are found explicitly. For example, such are the polynomials composed as a superposition of polynomials $q_j(z)$, $j = 1, 2, \ldots, k$,

$$Q_m(z) = q_1(q_2(\ldots q_k(z))), \tag{11.57}$$

where each $q_j(z)$ is a polynomial of one of the types: (a) a polynomial of power not higher than 4; (b) a polynomial of power equal to z^n; or (c) Chebyshev polynomial $T_n(z)$.

Let $\lambda > 0$, $\rho = 1$, $\theta = 1 + \varepsilon/n^2$, $\varepsilon > 0$, $t = u + iv = 1 - 2z/\lambda$. The function t maps the segment $[0, \lambda]$ to [-1,1].

Example 11.1.

$$Q_n(z) = T_n(\theta t)/T_n(\theta). \tag{11.58}$$

Figure 12 shows the boundary $\partial\Omega$ of the domain Ω_1 with inscribed ellipse focused in the point $(\lambda, 0)$ for $n = 8$, $\lambda = 2$, and $\varepsilon = 0.3$.

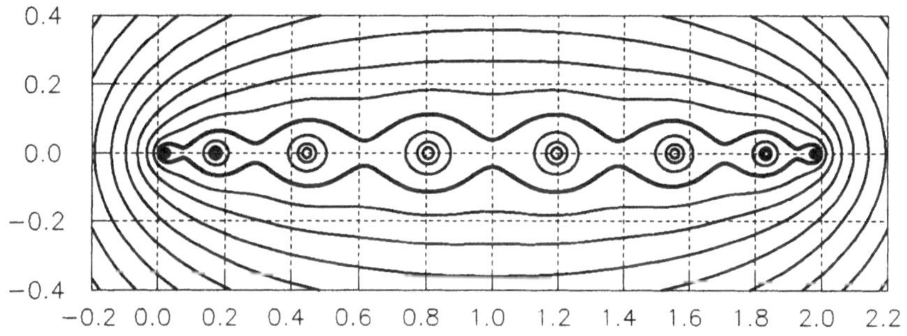

Figure 12

Example 11.2. Let $w = (2t^2 - 1 - b^2)/(1 - b^2)$. The function w maps onto [-1,1]: for $0 < b < 1$ a pair of segments $[-1, -b] \cup [b, 1]$; for $b = i\beta$, $\beta > 0$ the cross

$$\{-1 \le u \le 1,\ v = 0\} \cup \{u = 0,\ -i\beta \le v \le i\beta\}.$$

Let $Q_{2n}(z) = T_n(\theta w)/T_n(\theta)$. If $b = i\beta$ is imaginary, then the bound $\partial\Omega_1$ of the domain Ω_1 encloses a cross centered in the point $(\lambda/2, 0)$: $\{0 \le x \le \lambda,\ y = 0\} \cup \{x = \lambda/2,\ 0 \le |y| \le b\lambda/2\}$. (See Figure 13 for $n = 9$, $\lambda = 1$, $\beta = 2$, $\varepsilon = 0.05$.)

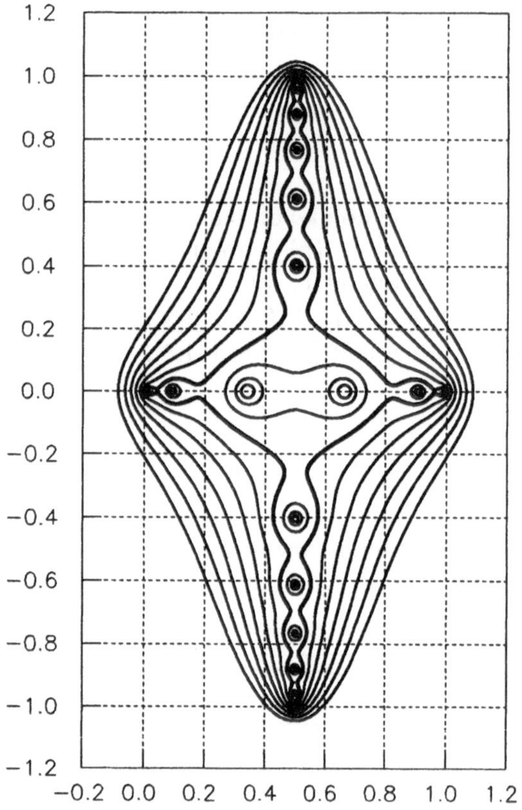

Figure 13

If $0 < b < 1$, then the bound $\partial\Omega_1$ will enclose two segments of a real axis of the same length

$$\{y = 0\, x \in [0, \lambda(1-b)/2] \cup [\lambda(1+b)/2, \lambda]\}.$$

(See Figure 14 for $n = 2$, $\lambda = 2$, $b = 0,4$, and $\varepsilon = 0.05$.)

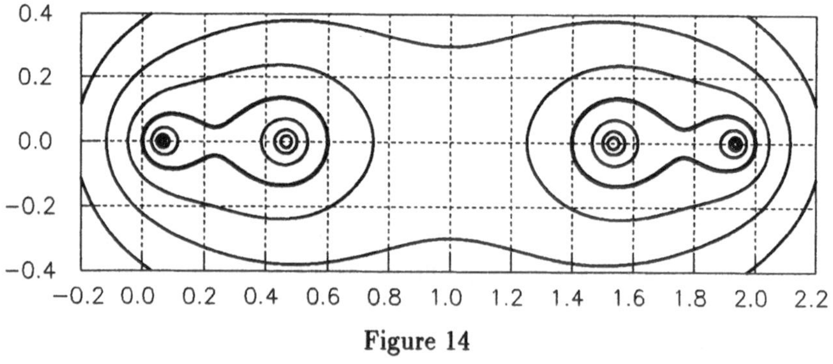

Figure 14

6. Solution of Problem V on the best approximation method with the Lagrange interpolation polynomial

We have obtained a solution of the extremal Problem IV in Example 7 of Section 3. Let us pass now to the study of extremal Problem V (3.20) on the best approximation method in $C([-1,1],w)$ on the class C of functions satisfying condition (3.14). Operator P of Problem IV depends on the numbers $x_1, x_2, \ldots, x_n \in [-1,1]$; let us regard these numbers as parameters for the optimization of the method. Varying these parameters, we obtain a family of operators that are denoted by $\mathcal{P}$. Then by (3.18) we have

$$S(C,\mathcal{P},C([-1,1],w)) = \inf_{x_1,x_2,\ldots,x_n} G(C,P,C([-1,1],w))$$

$$= \frac{R}{n!} \inf_{x_1,x_2,\ldots,x_n} \max_{-1\le x\le 1} |\omega_n(x)w(x)|. \quad (11.59)$$

This formula shows that in $C([-1,1],w)$ for n times continuously differentiable on [-1,1] functions the roots of extremal polynomial (11.35) of power n are the best nodes for interpolation with the Lagrange polynomial of power $(n-1)$. Let $w(x)=1$, according to Problem I (b) from Section 10, and let inf in (11.59) be attained for $\omega_n(x) = 2^{1-n}T_n(x)$ equal to 2^{1-n}; that is,

$$S(C,\mathcal{P},C[-1,1]) = \frac{R}{2^{n-1}n!}. \quad (11.60)$$

In this case x_k are defined by (10.12) and the polynomial (3.15) has the form

$$L_{n-1}(x) = \frac{1}{n}\sum_{k=1}^{n}(-1)^{k-1}f(x_k)\sqrt{1-x_k^2}\,\frac{T_n(x)}{x-x_k}. \quad (11.61)$$

Remark 1. The more general case, implying corresponding problems in the space $C([a,b],w)$, is reduced to the preceding approach by the substitution of a variable in (10.34).

Chapter 2

Linear Operators and Functionals

We defined the notion of operators in Section 1 of Chapter 1 and presented the definition of their continuity. Now we study an important class of these operators and functionals, namely, linear operators and functionals.

§ 1. Linear Operators in Banach Spaces

Definition of linear operator, its domain of definition, and range of values. Extension of an operator. Examples. Continuity, boundedness, norm of linear operator. Connection between continuity and boundedness. Estimates of norms for a sum and product of operators. On the extension of linear operator with respect to continuity.

1. Linear operator, domain of definition, range of values. Extension of an operator

Let $\mathbf{X}, \mathbf{Y}$ be two Banach spaces and let there exist an operator (mapping) $A : \mathbf{X} \to \mathbf{Y}$ that sets an element $y \in \mathbf{Y}$ into correspondence to an element $x \in \mathbf{X}$. Let the operator be defined on a set $D(A) \subset \mathbf{X}$ called the *domain of definition of the operator* A. A set of elements of the kind $R(A) = \{y \in \mathbf{Y},\ y = Ax,\ x \in D(A)\}$ is called the *range of values of the operator* A. If

$$y = Ax, \qquad x \in D(A), \qquad y \in R(A) \tag{1.1}$$

then we say that the element y is an *image* of the element x, and the element x is a *preimage* of the element y. Thus $R(A)$ is an image of $D(A)$ or, in short and conventionally, $R(A) = AD(A)$.

An operator $A : \mathbf{X} \to \mathbf{Y}$ with a domain of definition $D(A)$ is termed *linear* if:

(1) $D(A)$ is linear space; or

(2) $A(\lambda_1 x_1 + \lambda_2 x_2) = \lambda_1 A x_1 + \lambda_2 A x_2$ for $\forall\, x_1, x_2 \in D(A)$ and any scalars λ_1, λ_2.

If $Ax = 0$ for $\forall\, x \in \mathbf{X}$, then we term the operator A a *zero* operator.

Lemma 1. *The domain of definition of any linear operator is a linear space.*

□ Let y_1, $y_2 \in R(A)$, and λ_1, λ_2 be scalars. Take the elements $x_i \in D(A)$ that are preimages of elements y_i; that is, $Ax_i = y_i$, $i = 1, 2$. Use the second property of the linearity definition for A to obtain

$$\lambda_1 y_1 + \lambda_2 y_2 = \lambda_1 A x_1 + \lambda_2 A x_2 = A(\lambda_1 x_1 + \lambda_2 x_2).$$

Thus, by the first property, the element $\lambda_1 x_1 + \lambda_2 x_2$ is a preimage of the element $\lambda_1 y_1 + \lambda_2 y_2$, which means that the latter belongs to $R(A)$. ■

Linear operator A is called an *extension* of linear operator $B : \mathbf{X} \to \mathbf{Y}$ if $D(A) \supset D(B)$ and $Ax = Bx$ for $\forall\, x \in D(B)$. Two cases are usually considered: $D(A) = \mathbf{X}$ and $\overline{D(A)} = \mathbf{X}$. They say in the second case that the operator A is assigned on everywhere dense in $\mathbf{X}$ set $D(A)$.

2. Examples

1.1. In the space $\mathbf{R}^n$, equality (1.1), where $A = \{a_{ij}\}$, $i, j = \overline{1, n}$ is a square matrix of an order n, and x, y are vector-columns from $\mathbf{R}^n$, assigns an operator A.

1.2. The formal algebraic expression $y_i = \sum\limits_{j=1}^{\infty} a_{ij} x_j$, $i = 1, 2 \ldots$, where x and y are vectors of infinite order, may define, under some restrictions for the matrix $\{a_{ij}\}$, linear operators A in normed spaces of sequences. For example:

$$\text{if} \quad \sup_i \sum_{j=1}^{\infty} |a_{ij}| < \infty, \quad \text{then} \quad A : m \to m$$

$$\text{if} \quad \sum_{i,j=1}^{\infty} |a_{ij}|^2 < \infty, \quad \text{then} \quad A : l_2 \to l_2.$$

1.3. The integral expression $y(x) = \int_a^b K(x,s)x(s)ds$, where $K(x,s) \in C([a,b] \times [a,b])$, may define various integral operators $A : \mathbf{X} \to \mathbf{Y}$; for instance, $\mathbf{X} = C[a,b]$, $\mathbf{Y} = C[a,b]$ or $\mathbf{X} = L_2(a,b)$, $\mathbf{Y} = L_2(a,b)$, and so on.

1.4. The linear differential operators, defined by differential expression $Au = \sum_{0 \le |\alpha| \le l} a_\alpha(x) D^\alpha u$ with coefficients $a_\alpha(x) \in C(\overline{\Omega})$, $\overline{\Omega} \in \mathbf{R}^n$, map $C^l(\overline{\Omega})$ onto $C(\overline{\Omega})$.

3. Continuity, norm of an operator, boundedness. Connection between continuity and boundedness

Let $D(A) = \mathbf{X}$. Recall that an operator A is referred to as *continuous in a point* $x_0 \in \mathbf{X}$ if $Ax_n \to Ax_0$ for $x_n \to x_0$. The following theorem happens to be valid for linear operators.

Theorem 1. Let linear operator A be assigned everywhere in a Banach space $\mathbf{X}$ with values in a Banach space $\mathbf{Y}$ and let it be continuous in a point $0 \in \mathbf{X}$; then A is continuous in any point $x_0 \in \mathbf{X}$.

□ Since $Ax_n - Ax_0 = A(x_n - x_0)$, and if $x_n \to x_0$, then $z_n = x_n - x_0 \to 0$ for $n \to \infty$. By continuity in zero $Az_n \to 0$ and then $Ax_n - Ax_0 \to 0$. ■

Therefore, linear operator A is termed *continuous* if it is continuous in the point $x = 0$. The linear operator is referred to as *bounded* if the following value is finite

$$||A|| = \sup_{x \in \mathbf{X}} \frac{||Ax||_{\mathbf{Y}}}{||x||_{\mathbf{X}}} = \sup_{||x||_{\mathbf{X}}=1} ||Ax||_{\mathbf{Y}}. \tag{1.2}$$

The value $||A||$ is called a *norm* of operator A. Norms for Ax and x in (1.2) are taken in different spaces. It follows from (1.2) that

$$||Ax|| \le ||A|| \cdot ||x||. \tag{1.3}$$

(As a rule, we omit hereafter the norms' indices.) Let us show that the notions of continuous and bounded operators are equivalent.

Theorem 2. *Let $A : \mathbf{X} \to Y$ be a linear operator and* $\mathbf{X}$, $\mathbf{Y}$ *be Banach spaces,* $D(A) = \mathbf{X}$. *For A to be continuous, it is necessary and sufficient that it is bounded.*

□ *Necessity.* Let A be continuous but not bounded; that is, for $\forall\, n \geq 1$ the elements $x_n \in \mathbf{X}$ with $\|x_n\| = 1$ will be found such that $\|Ax_n\| \geq n$. Take new elements $x'_n = (1/n)x_n \in \mathbf{X}$ for which $\|x'_n\| = (1/n) \to 0$ and $Ax'_n \to 0$ for $n \to \infty$ by continuity of A. On the other hand, $\|Ax'_n\| = (1/n)\|Ax_n\| \geq 1$. The contradiction so obtained proves the first part of the theorem.

Sufficiency. Let A be bounded, by inequality (1.3) $Ax \to 0$ for $x \to 0$; that is, A is continuous in the point 0 and, therefore, is continuous everywhere. ■

4. Estimates of norms for a sum and product of operators. On the extension of a linear operator with respect to continuity

Let A and B be linear bounded operators, $D(A) = D(B) = \mathbf{X}$, $R(A), R(B) \subset \mathbf{Y}$. An operator defined for $\forall\, x \in \mathbf{X}$ by the equality

$$Cx = Ax + Bx$$

is called the *sum of operators* $C = A + B : \mathbf{X} \to \mathbf{Y}$. Let us estimate a norm of the operator C through the norms of operators A and B. We have

$$\|Cx\| = \|Ax + Bx\| \leq \|Ax\| + \|Bx\| \leq (\|A\| + \|B\|)\,\|x\|.$$

We obtain, taking into account the definition of a norm of operator (1.2), that

$$\|C\| \leq \|A\| + \|B\|. \tag{1.4}$$

Let linear operators $A : \mathbf{X} \to \mathbf{Y}$, $B : \mathbf{Y} \to \mathbf{Z}$ be such that $D(A) = \mathbf{X}$, $R(A) \subset D(B) = \mathbf{Y}$, $R(B) \subset \mathbf{Z}$ where $\mathbf{X}$, $\mathbf{Y}$, $\mathbf{Z}$ are Banach spaces. Then on the set $D(A)$ the operator $C = B{\cdot}A : \mathbf{X} \to \mathbf{Z}$ is defined that is called a *product of operators* B and A determined by the equality for $\forall\, x \in \mathbf{X}$

$$Cx = BAx = B(Ax). \tag{1.5}$$

Let us estimate a norm of this operator. We have

$$||Cx||_{\mathbf{Z}} = ||BAx||_{\mathbf{Z}} \leq ||B||\,||Ax||_{\mathbf{Y}} \leq ||B||\,||A||||x||_{\mathbf{X}}.$$

Recall the definition of a norm of operator (1.2) to obtain

$$||C|| \leq ||B|| \cdot ||A||. \tag{1.6}$$

Let $\overline{D(A)} = \mathbf{X}$ and let

$$||A|| = \sup_{\substack{x \in D(A) \\ ||x||=1}} ||Ax|| < \infty. \tag{1.7}$$

Operators A, satisfying condition (1.7), are called operators *bounded* on $D(A)$.

The following statement (we omit a proof here) is valid.

Theorem 3. (on the extension of a linear operator with respect to continuity). *Let* $\mathbf{X}$ *be normed space and* $\mathbf{Y}$ *Banach space and let* A *be a linear operator with* $D(A) \subset \mathbf{X}$, $R(A) \subset \mathbf{Y}$, *where* $\overline{D}(A) = \mathbf{X}$ *and the operator* A *is bounded on* $D(A)$ *in the sense of inequality* (1.7). *Then there exists a linear bounded operator* $\hat{A}$ *(an extension of the operator* A*) such that:* (1) $\hat{A}x = Ax$ *for* $\forall\, x \in D(A)$ and (2) $||\hat{A}|| = ||A||$.

There is a more complicated case when $\overline{D}(A) = \mathbf{X}$, but inequality (1.7) is not satisfied.

§ 2. Spaces of Linear Operators

Banach space of operators $\mathcal{L}(\mathbf{X}, \mathbf{Y})$. Strong convergence of operators and convergence with respect to norm; relationship between them. Series in $\mathcal{L}(\mathbf{X}, \mathbf{Y})$. Space $\mathcal{L}(\mathbf{X})$. Commutativity. Operator series, functions of operators from $\mathcal{L}(\mathbf{X})$. Neumann series. Projection operators.

1. Banach space of operators $\mathcal{L}(\mathbf{X}, \mathbf{Y})$

Let $\mathbf{X}, \mathbf{Y}$ be Banach spaces and let $A, B, C \ldots$ be sets of linear continuous operators defined everywhere in $\mathbf{X}$ with values in $\mathbf{Y}$. Define summing and multiplication scalar operations on this set of operators with the formulas

$$\begin{gathered}(A + B)x = Ax + Bx \\ (\lambda A)x = \lambda Ax.\end{gathered} \tag{2.1}$$

It is easy to show that $A+B$ and λA are linear continuous operators. Thus, the set of all linear continuous operators is a linear set, since it satisfies all axioms of linear space. Define a norm for elements of this space formula (1.2). Then $\|\lambda A\| = |\lambda| \cdot \|A\|$ and the triangle inequality (1.4) is valid for the operators A, B, $A+B$ and we arrive at the following theorem.

Theorem 1. A set of linear continuous operators defined everywhere in Banach space $\mathbf{X}$ *with values in Banach space* $\mathbf{Y}$ *is normed space.*

It is possible to show, as a matter of fact, that this space denoted by $\mathcal{L}(\mathbf{X}, \mathbf{Y})$ is a Banach one.

The convergence of a sequence A_n of operators in this space (called the *convergence with respect to the norm* or *uniform convergence*) to an operator A is determined by the condition:

$$\|A_n - A\| \to 0 \quad \text{for} \quad n \to \infty.$$

It is possible to define one more type of the convergence of operator sequence $\{A_n\}$ from $\mathcal{L}(\mathbf{X}, \mathbf{Y})$, so-called strong convergence: we say that $\{A_n\}$ *converges strongly* to an operator $A \in \mathcal{L}(\mathbf{X}, \mathbf{Y})$ if for $\forall x \in \mathbf{X}$ $\|A_n x - Ax\| \to 0$ for $n \to \infty$.

Lemma 1. If $A_n \to A$, $n \to \infty$ *with respect to a norm, then* $A_n \to A$, $n \to \infty$ *strongly.*

□ The statement follows from the estimate:

$$\|A_n x - Ax\| \leq \|A_n - A\| \, \|x\|. \blacksquare$$

Strong convergence of operators does not imply their convergence with respect to a norm; we see it, having studied properties of projection operators in item 3 of this section.

According to the general definition of the convergence for series (see Section 6 of Chapter 1) a series

$$\sum_{k=1}^{\infty} A_k, \qquad A_k \in \mathcal{L}(\mathbf{X}, \mathbf{Y}) \tag{2.2}$$

converges with respect to a norm if the sequence of its partial sums $S_n = \sum_{k=1}^{n} A_k$ converges with respect to a norm. We say that the series (2.2) *converges absolutely* if the number series $\sum_{k=1}^{\infty} \|A_k\|$ converges.

Lemma 2. *If the series* (2.2) *converges absolutely, then it also converges with respect to a norm.*

□ We have

$$\|S_{n+p} - S_n\| = \|\sum_{k=n+1}^{n+p} A_k\| \leq \sum_{k=n+1}^{n+p} \|A_k\|. \blacksquare$$

2. The space $\mathcal{L}(\mathbf{X})$. Commutativity. Operator power series and functions of operators. Neumann series

The space $\mathcal{L}(\mathbf{X}, \mathbf{X}) = \mathcal{L}(\mathbf{X})$ is met in applications most frequently. It is possible to introduce in the space $\mathcal{L}(\mathbf{X})$ one more operation: the multiplication of operators $A, B \in \mathcal{L}(\mathbf{X})$ with formula (1.5); obviously, $BA \in \mathcal{L}(\mathbf{X})$. Generally speaking, $BA \neq AB$. Two operators $A, B \in \mathcal{L}(\mathbf{X})$ are referred to as *commutative* or permutational if $BA = AB$. Estimates (1.4) and (1.6), proven before, stay valid for $A, B \in \mathcal{L}(\mathbf{X})$:

$$\|AB\| \leq \|A\| \cdot \|B\|, \quad \|A + B\| \leq \|A\| + \|B\|. \tag{2.3}$$

In $\mathcal{L}(\mathbf{X})$ a *unit* (*identical*) operator I is defined, determined by the equality $Ix = x$ for $\forall\, x \in \mathbf{X}$, as well as the power A^k (k is natural) of an operator A: $A^2 = A \cdot A$; $A^3 = A \cdot A^2, \ldots, A^k = A \cdot A^{k-1}, \ldots$; $A^0 = I$. Then $\|A^k\| \leq \|A\|^k$. This makes it possible to introduce polynomials of operators

$$P_N(A) = \sum_{k=0}^{N} a_k A^k \tag{2.4}$$

and functions of operators. Let $\varphi(\lambda) = \sum_{k=0}^{\infty} a_k \lambda^k$ be an analytic in a circle $|\lambda| < R$ function of complex variable λ, $A \in \mathcal{L}(\mathbf{X})$ and let $\|A\| < R$. Then define a function $\varphi(A)$ of operator A by the formula

$$\varphi(A) = \sum_{k=0}^{\infty} a_k A^k, \tag{2.5}$$

for example,

$$e^A = \sum_{k=0}^{\infty} \frac{A^k}{k!} \tag{2.6}$$

or

$$\sum_{k=0}^{\infty} A^k \tag{2.7}$$

for $||A|| < 1$. Series (2.7) is called a *Neumann series.*

An operator $P \in \mathcal{L}(X)$ such that $P^2 = P$ is called a *projection operator.* For instance, the operator $L_{n-1}f = L_{n-1}(x)$, where $L_{n-1}(x)$ is the Lagrange polynomial for the function $f \in C[a,b]$ [see (3.15), Chapter 1], is a projection operator. If P is a projection operator, then operator $\tilde{P} = I - P$ is a projection operator too, since $\tilde{P}^2 = I - P = \tilde{P}$.

3. Operators of orthogonal projection in a Hilbert space

Let $\mathbf{X} = \mathbf{Y} = H$ be a separable Hilbert space and let L be its subspace. Then according to Theorem 3 from Section 8 of Chapter 1 any element of $x \in H$ can be uniquely represented in the form

$$x = y + z, \tag{2.8}$$

where $y \in L,\ z \in L^{\perp}$. Define operator P_L by a formula: for $\forall\, x \in H$

$$y = P_L x. \tag{2.9}$$

□ It is obvious that it is a linear operator and $P^2 = P$ and $||P_L|| = 1$, since $||x||^2 = ||y||^2 + ||z||^2$ which means that $||y||^2 = ||P_L x||^2 = ||x||^2 - ||z||^2 \leq ||x||^2$ where the equality is attained for $x \in L$. ■

An operator P_L defined by relationships (2.8) and (2.9) is called the *orthogonal projection operator* of a space H onto subspace L, in short, the *orthoprojector* of H on L.

Let $\{e_n\}_1^\infty$ be the orthonormed basis in H, L_N – finite-dimensional subspace in H, spanned over the first N elements of the basis $\{e_n\}_1^\infty$, and let P_{L_N} be an orthogonal projection operator onto L_N. Then for $\forall\, x \in H$ there exists an expansion

$$x = \sum_{n=1}^{\infty} a_n e_n, \qquad a_n = (x, e_n)$$

and the operator P_{L_N} has the representation

$$P_{L_N} x = \sum_{n=1}^{N} a_n e_n. \tag{2.10}$$

It is obvious that $P_{L_N} \to I$ strongly since

$$\|P_{L_N}x - Ix\|^2 = \|\sum_{n=N+1}^{\infty} a_n e_n\|^2 = \sum_{n=N+1}^{\infty} a_n^2 \to 0, \quad n \to \infty$$

though

$$\|P_{L_N} - I\| = \sup_{\|x\|=1} \|(P_{L_N} - I)x\| \geq \|(I - P_{L_N})e_{N+1}\| = \|e_{N+1}\| = 1.$$

Thus, a strongly converging sequence of orthoprojectors P_{L_N} does not converge with respect to a norm.

§ 3. Inverse Operators. Linear Operator Equations. Condition Measure of Operator

Main notions, theorems on the existence of inverse operator. Linear operator equations of the first kind and their solution with inverse operator. Condition measure of operator and its application to the estimation of approximate solution.

1. Main notions, theorems on the existence of inverse operator

While studying (Section 5 of Chapter 1) linear mappings (operators) $A : \bar{\mathbf{X}} \to \bar{\mathbf{Y}}$ where $\bar{\mathbf{X}}$, $\bar{\mathbf{Y}}$ are linear spaces for $D(A) = \bar{\mathbf{X}}$, $R(A) = \bar{\mathbf{Y}}$ we proved Lemma 1 stating that an operator A maps $D(A)$ onto $R(A)$ in a one-to-one way, iff its kernel $N(A) = \ker A$ consists of a zero element. There exists, under conditions of this lemma, *inverse operator* A^{-1}, mapping $R(A)$ in a one-to-one way onto $D(A)$. Let us prove that the operator A^{-1} is linear.

□ Let $y_1, y_2 \in R(A)$ and let $x_i = A^{-1}y_i$, $i = 1, 2$, be their preimages. Since $R(A)$ is linear space (see Lemma 1 from Section 1 of this chapter), then for any scalars λ_1, λ_2 we have

$$A(\lambda_1 x_1 + \lambda_2 x_2) = \lambda_1 y_1 + \lambda_2 y_2 \in R(A)$$

or

$$\lambda_1 A^{-1} y_1 + \lambda_2 A^{-1} y_2 = \lambda_1 x_1 + \lambda_2 x_2 = A^{-1}(\lambda_1 y_1 + \lambda_2 y_2). \blacksquare$$

So, if an operator $A : \mathbf{X} \to \mathbf{Y}$ is assigned and if there exists an operator $A^{-1} : \mathbf{Y} \to \mathbf{X}$ defined on $R(A)$ and taking values in $D(A)$ such that

$$A^{-1}Ax = x \quad \text{for} \quad \forall\, x \in D(A)$$

and (3.1)

$$AA^{-1}y = y \quad \text{for} \quad \forall\, y \in R(A)$$

then operators A and A^{-1} are termed *mutually inverse* and operator A^{-1} is referred to as *inverse*. Then $(A^{-1})^{-1} = A$. In the space $\mathcal{L}(X)$ the operators A, $A^{-1} \in \mathcal{L}(X)$ are characterized by the equalities

$$A^{-1}A = I, \qquad AA^{-1} = I. \tag{3.2}$$

If $B, B^{-1} \in \mathcal{L}(\mathbf{X})$ too, then

$$(AB)^{-1} = B^{-1}A^{-1}.$$

Theorem 1. *An operator A^{-1} exists and at the same time is bounded on $R(A)$ iff there exists some constant $m > 0$ such that for any $x \in D(A)$ the following equality is valid*

$$\|Ax\| \geq m\|x\|. \tag{3.3}$$

□ *Necessity.* Let A^{-1} exist and be bounded on $R(A) = D(A^{-1})$. It means that there exists $C > 0$ such that $\|A^{-1}y\| \leq\, \leq C\|y\|$ for $\forall\, y \in R(A)$. Having set $y = Ax$ we obtain (3.3) with $C = m^{-1}$.

Sufficiency. If (3.3) is satisfied, then by equality $Ax = 0$ and by (3.3) $x = 0$; that is, $N(A) = \ker A = \{0\}$. By Lemma 1 from Section 5 of Chapter 1, there exists A^{-1} mapping $R(A)$ in a one-to-one way onto $D(A)$. Having set in (3.3) $x = A^{-1}y$, we obtain $\|A^{-1}y\| \leq m^{-1}\|y\|$ for $\forall\, y \in R(A)$; that is, A^{-1} is bounded on $R(A)$ and $\|A^{-1}\| \leq m^{-1}$, if $R(A) = \mathbf{Y}$. ■

An operator inverse to a linear bounded one, is not necessarily a linear bounded operator. We say that linear operator $A : \mathbf{X} \to \mathbf{Y}$ is *continuously invertible* if $R(A) = \mathbf{Y}$, the operator A is invertible, and $A^{-1} \in \mathcal{L}(\mathbf{Y}, \mathbf{X})$, that is, bounded.

Corollary 1. *An operator A is continuously invertible iff $R(A) = \mathbf{Y}$ and for some constant $m > 0$ and $\forall\, x \in D(A)$ inequality (3.3) is satisfied with $\|A^{-1}\| \leq m^{-1}$.*

The Banach theorem is valid, saying that *if* $A \in \mathcal{L}(\mathbf{X}, \mathbf{Y})$ *where* $\mathbf{X}$ *and* $\mathbf{Y}$ *are Banach,* $R(A) = \mathbf{Y}$*, and* A *is invertible, then* A *is continuously invertible.* We omit its proof here.

Other sufficient conditions for the existence of an inverse operator are discussed in Section 4.

If operator A^{-1} satisfies just the first (or the second) of conditions (3.1) and (3.2), then it is termed *left* (respectively, *right*) *inverse* to the operator A and denoted by A_l^{-1} (respectively, A_r^{-1}).

2. Linear operator equations of the first kind and their solution with inverse operator

Linear algebraic equations, linear integral equations, and problems for differential equations (ordinary or with partial derivatives) can be written as a *linear operator equation of the* 1*st kind* as related to an unknown $x \in \mathbf{X}$:

$$Ax = y \tag{3.4}$$

with linear operator A and given element $y \in \mathbf{Y}$. The questions arise, first of all, on the existence, uniqueness, and correct solvability of problem (3.4).

So, take Equation (3.4) and assume that $y \in R(A)$ and the operator A has the inverse one A^{-1}. Let us set

$$x = A^{-1}y \tag{3.5}$$

and substitute this value into equality (3.4) to obtain the identity $AA^{-1}y = y$; that is, $y = y$. Therefore, (3.5) is a solution to Equation (3.4). Let us take that there exists another solution x_1 to Equation (3.4); that is, $Ax_1 = y$. Having applied the operator A^{-1} to both sides of this equality we obtain $x_1 = A^{-1}y = x$. Therefore, solution (3.5) is unique. If A is continuously invertible, then Equation (3.4) has a unique solution for any right-hand part $y \in \mathbf{Y}$. If, at the same time, $\bar{x} = A^{-1}\bar{y}$ (a solution of the same equation with another right-hand part $\bar{y}$), then $\|x - \bar{x}\| \leq \|A^{-1}\| \, \|y - \bar{y}\|$. It means that a small change in the right-hand part y in (3.4) implies a small change in the solution; that is, as we say, problem (3.4) is *correctly resolved.*

If an operator A has just the right inverse operator A_r^{-1}, then it is easily seen that $x = A_r^{-1}y$ is a solution to Equation (3.4), though its uniqueness should be checked. If an operator A has a left inverse operator and Equation (3.4) has a solution x, then applying operator A_l^{-1} from the left to both parts of equality (3.4) we obtain $x = A_l^{-1}y$; that is, the solution is unique, though its existence should be checked.

3. Condition measure of an operator and its application to the estimation of approximate solution of a linear operator equation with continuously invertible operator

Let $A,\ A^{-1} \in \mathcal{L}(\mathbf{X})$, and let $\mathbf{X}$ be Banach space. We call a number

$$K(A) = \|A\| \cdot \|A^{-1}\| \tag{3.6}$$

the *condition number* of the operator A. It is obvious that $K(A) \geq 1$.

□ In fact, for $x \neq 0$ we have $x = AA^{-1}x$ and by inequality (1.6)

$$\|x\| \leq \|A\|\|A^{-1}\|\|x\| = K(A)\|x\|$$

obtain the desired inequality. ■

Let u be an approximate solution to equation (3.4), $\varepsilon = x - u$ be an *error*, and $r = y - Au$ be its *discrepancy*. Then we have the following relationships:

$$x = A^{-1}y, \qquad A\varepsilon = r, \qquad \varepsilon = A^{-1}r \tag{3.7}$$

from them and by Equation (3.4) we obtain four inequalities:

$$\begin{gathered} \|y\| \leq \|A\|\|x\|, \qquad \|x\| \leq \|A^{-1}\|\|y\| \\ \|r\| \leq \|A\|\|\varepsilon\|, \qquad \|\varepsilon\| \leq \|A^{-1}\|\|r\|. \end{gathered} \tag{3.8}$$

Let us show that the following inequality is valid for a *relative error* $\|\varepsilon\|/\|x\|$:

$$K^{-1}(A)\frac{\|r\|}{\|y\|} \leq \frac{\|\varepsilon\|}{\|x\|} \leq K(A)\frac{\|r\|}{\|y\|}. \tag{3.9}$$

This inequality contains just $K(A)$, the norm of the right-hand part of (3.4), and the norm of discrepancy. In fact, estimating $\|\varepsilon\|/\|x\|$ from above and using the first and the fourth inequalities from (3.8), we obtain

$$\frac{\|\varepsilon\|}{\|x\|} \leq \frac{\|A^{-1}\|\|r\|}{\|A\|^{-1}\|y\|} = K(A)\frac{\|r\|}{\|y\|}.$$

Similarly, using the second and the third inequalities from (3.8), we obtain

$$\frac{\|\varepsilon\|}{\|x\|} \geq \frac{\|A\|^{-1}\|r\|}{\|A^{-1}\|\|y\|} = K^{-1}(A)\frac{\|r\|}{\|y\|}.$$

Inequality (3.9) is proven; it can be useful for the estimation of accuracy for various approximate methods of the solution of Equation (3.4).

§4. Spectrum and Spectral Radius of Operator. Convergence Conditions for the Neumann Series. Perturbations Theorem

Resolvent of an operator, resolvent set, spectrum of an operator (discrete, continuous, residual). Eigenelements and eigenvalues. Examples. Spectral radius and a norm of operator. Nilpotent operators. Definition of selfadjoint operator and its spectral radius. Convergence and divergence conditions for the Neumann series. Theorem on existence of inverse operator $(I - A)^{-1}$. Perturbation theorem (existence of inverse operators).

1. Resolvent of an operator, resolvent set, spectrum of an operator (discrete, continuous, residual). Eigenelements and eigenvalues. Examples

Let $A \in \mathcal{L}(\mathbf{X})$, $\mathbf{X}$ be a *complex* Banach space and domain of definition $D(A)$ be everywhere dense in $\mathbf{X}$. Take an operator $A_\lambda = A - \lambda I$, where λ is a complex number (point of a complex domain). The so-called *spectral theory of operators* investigates the set of values of λ such that the operator A_λ has no inverse one, and studies the properties of an inverse operator to A_λ in cases when it exsists.

A point λ is termed the *regular* point of an operator A if the operator $A_\lambda = A - \lambda I$ is continuously reversible. A set of regular points of an operator A is called the *resolvent set* of operator A and is denoted by $\rho(A)$. If $\lambda \in \rho(A)$, then linear operator $R_\lambda(A) = (A - \lambda I)^{-1} \in \mathcal{L}(\mathbf{X})$ is called a *resolvent* of operator A. A complement to $\rho(A)$ (in a complex plane) is called the *spectrum* of operator A and is denoted by $\mathrm{Sp}(A)$.

Lemma 1. *The resolvent set $\rho(A)$ is always an open set.*

□ The lemma follows from a theorem to be proven in the following on perturbation of an operator. It is shown there that if the operator $A - \lambda_0 I$ is continuously reversible, then all operators $A - \lambda I$ close to it with respect to the norm possess the same property. ■

It follows from Lemma 1 that the spectrum of linear operator A is a closed set. We show in the following that it lies in the circle $|\lambda| \le \|A\|$ and, therefore, is a bounded set. A spectrum $\mathrm{Sp}(A)$ can be split into three mutually exclusive sets that correspond to three possible sets. Namely, the spectrum $\mathrm{Sp}(A)$ consists of:

(1) those λ for which the operator $A - \lambda I$ *is irreversible*; then $\ker A_\lambda = N(A_\lambda) \neq \{0\}$ and there exists an element $x \in \mathbf{X}$ such that

$$Ax = \lambda x. \tag{4.1}$$

An element x satisfying equation (4.1) is called an *eigenelement*, λ is called an *eigenvalue*, $N(A_\lambda)$ is called an *eigen subspace*, and (λ, x) is called an *eigenpair* of an operator A. A set of all eigenvalues of an operator A is referred to as a *point-wise* or *discrete spectrum* of operator A;

(2) those λ for which operator A_λ possesses an inverse operator with dense in $\mathbf{X}$ domain of definition, but operator A_λ^{-1} is not bounded; this set λ is called a *continuous spectrum*;

(3) those λ for which A_λ has an inverse operator, whose domain of definition is not everywhere dense in $\mathbf{X}$; this set λ is called a *residual spectrum.*

Examples

Example 4.1. If $\mathbf{X} = \mathbf{R}^n$, then any linear operator A has corresponding matrix $\{a_{ij}\}$. Eigenvalues of operator A are roots of the characteristic equation of the matrix $\{a_{ij}\} : \det(a_{ij} - \lambda\delta_{ij}) = 0$ in this case where $\det A$ denotes a determinant of the matrix A.

Example 4.2. Let $\mathbf{X} = L_2(0,1)$ and $Af(x) = xf(x)$ (A is a multiplication operator by an independent variable). The condition $(A - \lambda_0 I)f = 0$ means here that $(\lambda_0 - x)f(x) = 0$ almost everywhere and, therefore, $f(x) = 0$ almost everywhere too. Consequently, for $\forall\, \lambda \in [0,1]$ the operator $(A - \lambda_0 I)^{-1}$ exists. All functions $y(x) \in L_2(0,1)$, that turn identically to zero in some neighborhood of the point $x = \lambda_0$, belong to the domain of definition $D((A - \lambda_0 I)^{-1})$ and this set is dense in $L_2(0,1)$. But the operator $(A - \lambda_0 I)^{-1}$ is not bounded on an aggregate of such functions. Therefore, a spectrum of operator A is the segment [0,1] and it is continuous.

We notice for those who are familiar with the theory of generalized functions, that the value $\lambda = \lambda_0$ can be associated with a "generalized eigenfunction," which does not belong to the space $L_2(0,1)$, that is, delta-function $\delta(x - \lambda_0)$.

Example 4.3. Let $\mathbf{X} = l_2$ and let an operator A be defined by the formula

$$A(x_1, x_2, \ldots) = (0, x_1, x_2, \ldots) .$$

Then the number $\lambda = 0$ belongs to the residual spectrum of operator A, since the set $R(A)$ is not dense in l_2.

2. Spectral radius of linear operator $\mu(A)$. Formula for $\mu(A)$ through spectrum of an operator A. Relationship between $\mu(A)$ and $\|A\|$. Nilpotent operators. $\mu(A)$ of selfadjoint operator

A value

$$\mu(A) = \lim_{n\to\infty} \|A^n\|^{\frac{1}{n}} \tag{4.2}$$

is called the *spectral radius* of the operator A.

Theorem 1. Let $A \in \mathcal{L}(\mathbf{X})$; then there exists the finite limit (4.2) *and the relationship*

$$\mu(A) \leq \|A\| \tag{4.3}$$

is valid.

□ Let us prove the existence of a limit. Let $n = km+l$, $0 \leq l < m$, then

$$\|A^n\| = \|A^{km+l}\| \leq \|A^k\|^m \|A^l\|.$$

Since $m/n = 1/k - l/kn$, then

$$0 \leq \|A^n\|^{1/n} \leq \|A^k\|^{m/n}\|A^l\|^{1/n} = \|A^k\|^{1/k}\|A^k\|^{-l/(kn)}\|A^l\|^{1/n}.$$

Fix k and l in this inequality and turn m to infinity to obtain

$$0 \leq \overline{\lim_{n\to\infty}} \|A^n\|^{1/n} \leq \|A^k\|^{1/k} \leq \|A\| < \infty, \tag{4.4}$$

where $\overline{\lim}$ means upper bound; it always exists for an upper bounded set of numbers; the right-hand part of the inequality is obvious, since always $\|A^k\| \leq \|A\|^k$. Turn now $k \to \infty$ in (4.4) to obtain

$$\overline{\lim_{n\to\infty}} \|A^n\|^{1/n} \leq \underline{\lim_{k\to\infty}} \|A^k\|^{1/k} \leq \|A\|,$$

where lim means lower bound. So, a bound $\mu(A)$ exists, is finite, and inequality (4.3) is satisfied. ∎

The following formula is valid for the spectral radius

$$\mu(A) = \sup_{\lambda \in \mathrm{Sp}(A)} |\lambda|. \tag{4.5}$$

We prove it in Section 9 of this chapter for the case when an operator A is a selfadjoint one.

Corollary 1. $\mu(A^k) = \mu^k(A)$, $\mu(\alpha A) = |\alpha|\mu(A)$.

However, the spectral radius does not possess all the properties of a norm. In particular, $\mu(A) = 0$ does not imply $A = 0$.

Example 4.4. Let $\mathbf{X} = \mathbf{R}^2$. Let an operator A be a matrix of the kind $\begin{pmatrix} 0 & a \\ 0 & 0 \end{pmatrix}$ where $a \neq 0$ is a number. Then $\mu(A) = 0$, $\|A\| = |a| > 0$.

An operator A is termed *nilpotent* if $\mu(A) = 0$.

An operator $A \in \mathcal{L}(H)$ where H is a Hilbert space with an inner product $(\cdot,\cdot)$ is referred to as *selfadjoint* if for $\forall\, x, y \in H$ the following equality is valid

$$(Ax, y) = (x, Ay). \tag{4.6}$$

Theorem 2. *If A is a selfadjoint operator, then $\mu(A) = \|A\|$.*

□ We have $\|A^2\| \le \|A\|^2$, but $\|A\|^2 = \sup_{\|x\|=1} \|Ax\|^2 = \sup_{\|x\|=1} (Ax, Ax)$ $= \sup_{\|x\|=1} (A^2x, x) \le \sup_{\|x\|=1} \|A^2x\| = \|A^2\|$,; that is, $\|A^2\| = \|A\|^2$. It is proved in a similar way that $\|A^{2^m}\| = \|A\|^{2^m}$; therefore, $\mu(A) = \lim_{m\to\infty} \|A^{2^m}\|^{2^{-m}} = \|A\|$. ∎

3. Convergence and divergence conditions for the Neumann series. Existence theorem for inverse operator $(I - A)^{-1}$. Perturbation (existence) theorem for inverse operators

Let $A \in \mathcal{L}(\mathbf{X})$. Take the Neumann series for this operator

$$I + A + A^2 + \ldots + A^n + \ldots \tag{4.7}$$

and clarify the conditions when this series converges absolutely (we say converges, for the sake of simplicity).

Theorem 3. *If $\mu(A) < 1$, then the series* (4.7) *converges; if $\mu(A) > 1$, then the series* (4.7) *diverges.*

□ Let $\mu(A) = q < 1$. Then for $\forall \varepsilon > 0$ such that $\varepsilon < 1 - q$ the number n_0 will be found such that for $\forall n \geq n_0$ the inequality $\|A^n\|^{1/n} < q + \varepsilon < 1$ is valid, that is, $\|A^n\| < (q+\varepsilon)^n$ and then

$$\sum_{n=n_0}^{\infty} \|A^n\| \leq \sum_{n=n_0}^{\infty} (q+\varepsilon)^n$$

and the right-hand side of this inequality tends to zero for $n_0 \to \infty$, which means that the series (4.7) converges.

Now let $\mu(A) = q > 1$. Then for $\forall \varepsilon > 0$ such that $q - \varepsilon > 1$ the number n_1 will be found such that for $\forall n \geq n_1$ the inequality $\|A^n\|^{1/n} > q - \varepsilon > 1$ is valid, that is, $\|A^n\| > (q-\varepsilon)^n$, and then the nth member of the series $\sum_{n=1}^{\infty} \|A^n\|$ does not tend to zero for $n \to \infty$ which means that this series diverges. ■

Corollary 2. *If $\|A\| = q < 1$, then the Neumann series converges.*

□ The correctness of this corollary follows from inequality (4.3). ■

Corollary 3. *For the series* (4.7) *to converge, it is necessary and sufficient that for some k*

$$\|A^k\| < 1. \tag{4.8}$$

□ In fact, if the series converges, then $\|A^k\| \to 0$ and therefore (4.8) is valid for large enough k. On the contrary, if (4.8) is valid, then $\mu(A) < 1$ [see formulas (4.4) and (4.2)] and, consequently, the series (4.7) converges. ■

Theorem 4. *If $A \in \mathcal{L}(\mathbf{X})$ and $\|A\| = q < 1$, then the operator $I - A$ is continuously reversible with $\|(I-A)^{-1}\| \leq (1-q)^{-1}$.*

□ The theorem's conditions satisfy the conditions of Corollary 1, which means that the Neumann series converges. Denote by B a sum of this series:

$$B = \sum_{k=0}^{\infty} A^k. \tag{4.9}$$

It is easy to check that

$$B(I-A)=\sum_{k=0}^{\infty} A^k(I-A)=I \tag{4.10}$$

and

$$(I-A)B=\sum_{k=0}^{\infty}(I-A)A^k=I, \qquad (I-A)Bx=x; \tag{4.11}$$

that is,

$$B=(I-A)^{-1} \qquad \ker B=\{0\}. \tag{4.12}$$

Finally,

$$\|B\| \le \sum_{k=0}^{\infty} \|A^k\| \le \sum_{k=0}^{\infty} q^k=(1-q)^{-1}. \ \blacksquare$$

Remark. Since relationships (4.10) and (4.11) are always valid when the series (4.9) converges, then the operator $(I-A)^{-1}$ will exist when $\mu(A)<1$ or when $\|A^k\|<1$ for some $k \ge 1$ too.

We may show a domain now where $\mathrm{Sp}(A)$ lies.

Lemma 2. *Let $A \in \mathcal{L}(\mathbf{X})$. Then*

$$\{\lambda : |\lambda| > \|A\|\} \subset \rho(A).$$

□ Since $A-\lambda I=-\lambda(I-\lambda^{-1}A)$ and if $\|\lambda^{-1}A\|<1$, then $I-\lambda^{-1}A$ is continuously reversible by Theorem 4. ■

Corollary 4. $\mathrm{Sp}(A) \subset \{\lambda : |\lambda| \le \|A\|\}$.

Theorem 5 *on perturbations (on four balls). If A and $A^{-1} \in \mathcal{L}(\mathbf{X})$, then a set G of elements $\mathcal{L}(\mathbf{X})$, possesing an inverse in $\mathcal{L}(\mathbf{X})$, contains together with operators A and A^{-1} two balls:*

$$\begin{aligned} D_1&=\{B : \|A-B\| < \|A^{-1}\|^{-1}\} \\ D_2&=\{B : \|A^{-1}-B\| < \|A\|^{-1}\}. \end{aligned} \tag{4.13}$$

If an operator B lies in the ball D_1, then its inverse one can be represented with the series

$$B^{-1}=A^{-1}\sum_{n=0}^{\infty}[(A-B)A^{-1}]^n \tag{4.14}$$

or

$$B^{-1} = \sum_{n=0}^{\infty} [A^{-1}(A-B)]^n A^{-1} \tag{4.15}$$

with the inequality

$$\|B^{-1} - A^{-1}\| \le \frac{\|A^{-1}\|^2 \|A-B\|}{1 - \|A-B\|\,\|A^{-1}\|} \tag{4.16}$$

being valid; if $B_\varepsilon \in G$ *and* $\|B_\varepsilon - A\| \to 0$ *for* $\varepsilon \to 0$, *then* $\|A^{-1} - B_\varepsilon^{-1}\| \to 0$ *for* $\varepsilon \to 0$ *too. If operator* B *lies in the ball* D_2, *then its inverse one is representable with the series*

$$B^{-1} = A \sum_{n=0}^{\infty} [(A^{-1} - B)A]^n \tag{4.17}$$

or

$$B^{-1} = \sum_{n=0}^{\infty} [A(A^{-1} - B)]^n A \tag{4.18}$$

with the inequality

$$\|B^{-1} - A\| \le \frac{\|A\|^2 \|A^{-1} - B\|}{1 - \|A^{-1} - B\|\,\|A\|} \tag{4.19}$$

being valid; if $B_\varepsilon \in G$ *and* $\|B_\varepsilon - A^{-1}\| \to 0$ *for* $\varepsilon \to 0$, *then* $\|A - B_\varepsilon^{-1}\| \to 0$ *for* $\varepsilon \to 0$ *too* (Figure 15).

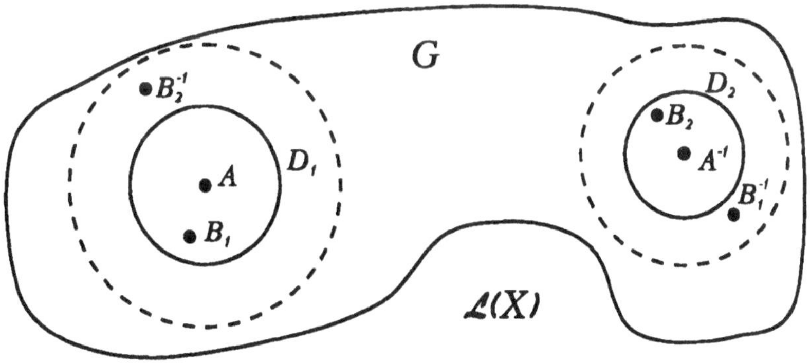

Figure 15

□ The proof of this theorem consists of two parts: for the ball D_1 and for the ball D_2. The proof for the ball D_2 is similar to the proof for the ball D_1, with the exception that operator A^{-1} should be substituted for operator A. So, let us prove the statement for the ball D_1.

Let $A \in G$ and $\|A - B\| < \|A^{-1}\|^{-1}$. Then, since $\|I - BA^{-1}\| = \|(A - B)A^{-1}\| < 1$ and $\|I - A^{-1}B\| = \|A^{-1}(A - B)\| < 1$ [see (4.13)], by Theorem 4 the operators $BA^{-1} = I - (I - BA^{-1})$ and $A^{-1}B = I - (I - A^{-1}B)$ have inverse ones that can be represented [see (4.13)], respectively, by the series

$$AB^{-1} = (BA^{-1})^{-1} = \sum_{n=0}^{\infty}(I - BA^{-1})^n = \sum_{n=0}^{\infty}[(A - B)A^{-1}]^n$$

$$B^{-1}A = (A^{-1}B)^{-1} = \sum_{n=0}^{\infty}(I - A^{-1}B)^n = \sum_{n=0}^{\infty}[A^{-1}(A - B)]^n$$

and, therefore, operator B has the inverse one B^{-1} represented by formulas (4.14) and (4.15). By (4.14) for $\|(A-B)A^{-1}\| < 1$ inequality (4.16) is satisfied

$$\|B^{-1}-A^{-1}\| = \|A^{-1}\sum_{n=1}^{\infty}[(A-B)A^{-1}]^n\| \leq \frac{\|A^{-1}\|^2\|A - B\|}{1-\|A-B\|\,\|A^{-1}\|} \tag{4.20}$$

which means that if $B_\varepsilon \in G$ and $\|A - B_\varepsilon\| \to 0$ for $\varepsilon \to 0$, then $\|B_\varepsilon^{-1} - A^{-1}\| \to 0$ too. The first part of the theorem is proven; as was mentioned, the second part is proven by the substitution of operator A^{-1} for operator A. ■

This important theorem is frequently used in the justification of numerical methods. Four balls in this theorem are the balls (4.13), (4.16), and (4.19).

§ 5. Uniform Boundedness Principle

Uniform boundedness, fixing, and singularities condensation principles. Banach–Steinhaus theorem. Application of the theorem to estimation of the Lagrange interpolation method and to representation of functions by the Fourier–Fejér and the de la Vallée-Poussin series in the space $C[-\pi, \pi]$.

Let us study properties of operator sequences $\{A_n\} \in \mathcal{L}(\mathbf{X}, \mathbf{Y})$ where $\mathbf{X}, \mathbf{Y}$ are Banach spaces.

1. Uniform boundedness, fixing, and singularities condensation principles. Banach–Steinhaus theorem

Lemma 1. *Let an operator sequence $\{A_n\} \in \mathcal{L}(\mathbf{X}, \mathbf{Y})$ and let there exist a constant $c > 0$ and closed ball $\overline{D}_r(x_0)$ $(r > 0)$ such that $\|A_n x\| \leq c$ for $\forall\, x \in \overline{D}_r(x_0)$ (i.e., the sequence $\{A_n x\}$ is uniformly bounded on $\overline{D}_r(x_0)$]. Then the sequence $\{\|A_n\|\}$ is bounded.*

□ Take $\forall\, x \in \mathbf{X}$, $x \neq 0$; then an element $x_0 + r\dfrac{x}{\|x\|} \in S_r(x_0)$ $\subset \overline{D}_r(x_0)$. Consequently,

$$c \geq \left\| A_n \left(\frac{rx}{\|x\|} + x_0 \right) \right\| = \left\| \frac{r}{\|x\|} A_n x + A_n x_0 \right\| \geq \frac{r}{\|x\|} \|A_n x\| - c$$

hence $\|A_n x\| \leq (2c/r)\|x\|$ and, therefore, $\|A_n\| \leq (2c/r)$. ■

Theorem 1. *(uniform boundedness principle). If $\{A_n x\}$ is bounded for each fixed $x \in \mathbf{X}$, then the sequence $\{\|A_n\|\}$ is bounded.*

□ Let us take that the theorem is incorrect. Then $\{\|A_n x\|\}$ is not bounded in any closed ball; otherwise by Lemma 1 $\{\|A_n\|\}$ would be bounded. Take some ball $\overline{D}_{r_0}(x_0)$, $r_0 > 0$, $x_0 \in \mathbf{X}$. Here $\{\|A_n\|\}$ is not bounded; therefore, an element $x_1 \in D_{r_0}(x_0)$ and number n_1 will be found such that $\|A_{n_1} x_1\| > 1$. By continuity of A_{n_1}, a ball $\overline{D}_{r_1}(x_1)$ $\subset D_{r_0}(x_0)$ will be found such that in it $\|A_{n_1} x\| > 1$ and $r_1 < r_0/2$. The sequence $\{\|A_n x\|\}$ in $\overline{D}_{r_1}(x_1)$ is also not bounded, so it is possible to find $x_2 \in D_{r_1}(x_1)$, $r_2 \leq r_1/2$, and $n_2 > n_1$ such that $\|A_{n_2} x_2\| > 2$ and so forth. As a result of such constructions, we find $\{x_k\}$, $\overline{D}_{r_k}(x_k)$ and $r_k \leq r_{k-1}/2$ with $x_k \in \overline{D}_{r_k}(x_k)$ and $\overline{D}_{r_0} \supset \overline{D}_{r_1} \supset \ldots \supset \overline{D}_{r_n} \supset \ldots$ and $\|A_{n_k} x\| > k$ on $\overline{D}_{r_k}$. By Theorem 3 on imbedded balls (see Section 1 of Chapter 1), their common point can be found: $\bar{x} \in \overline{D}_{r_k}$, $k = 1, 2, \ldots$. Then $\|A_{n_k} \bar{x}\| > k$; that is, $\{A_n \bar{x}\}$ is not bounded which contradicts the theorem's condition. ■

The reasoning of the preceding proof shows that two more theorems are valid.

Theorem 2. *(singularities fixing principle). If $\sup\limits_n \|A_n\| = \infty$, then an element $\bar{x} \in \mathbf{X}$ will be found such that $\sup\limits_n \|A_n \bar{x}\| = \infty$.*

It is possible to generalize this statement. Let $\{A_n^k\} \in \mathcal{L}(\mathbf{X}, \mathbf{Y})$, $k, n = 1, 2, \ldots$ with $\sup_n \|A_n^k\| = \infty$, $k = 1, 2, \ldots$.[1] Then there exists an element $x_0 \in \mathbf{X}$ such that

$$\sup_n \|A_n^k x_0\| = \infty, \qquad k = 1, 2, \ldots .$$

This statement presented with no proof is called the *singularities fixing principle.*

Theorem 3 (Banach–Steinhaus). Let $\{A_n\} \subset \mathcal{L}(\mathbf{X}, \mathbf{Y})$. *The following is necessary and sufficient for* $A_n \to A \in \mathcal{L}\mathbf{X}, \mathbf{Y})$, $n \to \infty$ *strongly:*

(1) $\{\|A_n\|\}$ *would be bounded;*

(2) $A_n x \to Ax$, $n \to \infty$, *for* $\forall\, x$ *from a set* $\mathbf{X}'$ *everywhere dense in* $\mathbf{X}$.

□ Necessity. It follows from the condition $A_n x \to Ax$, $n \to \infty$, $x \in \mathbf{X}$ that $\|A_n x\| \to \|Ax\|$, $n \to \infty$ and, therefore, $\{\|A_n x\|\}$ is bounded. By the uniform boundedness principle, $\{\|A_n\|\}$ is bounded. As $\mathbf{X}'$ one can take $\mathbf{X}$.

Sufficiency. Let $x \in \mathbf{X}$, but $x \notin \mathbf{X}'$. Let us set $\forall \varepsilon > 0$ and find $x' \in \mathbf{X}'$ such that $\|x - x'\| < \varepsilon$. Let $c = \sup_{n=0,1,\ldots} \|A_n\|$, where $A_0 = A$. Show that $A_n \to A$, $n \to \infty$ strongly:

$$\|A_n x - Ax\| = \|A_n(x - x') + (A_n x' - Ax') + A(x' - x)\|$$

$$\leq \|A_n\|\|x - x'\| + \|A_n x' - Ax'\| + \|A\|\|x' - x\|\| \leq 2c\varepsilon + \|A_n x' - Ax'\|$$

but $A_n x' \to Ax'$; therefore, there exists n_0 such that for $\forall\, n > n_0$ $\|A_n x' - Ax'\| < \varepsilon$. Then for $\forall\; n > n_0$ $\|A_n x - Ax\| < (2c + 1)\varepsilon$. ■

Many practical computing algorithms can be described by the same abstract scheme that contains the sequences of linear operators.

The questions of justification of such algorithms are usually reduced either to the establishment of convergence of a sequence of linear operators, or to the proof of boundedness for the norms of this sequence's operators. We use, in items 2 and 3 of this section, the

[1]Here k is index rather than power.

Banach–Steinhaus theorem for the investigation of properties of iteration polynomials, finite Fourier sums, and convergence of quadrature processes.

2. Application of the theorem to the estimation of the Lagrange interpolation method in the space $C[a,b]$. The Bernstein and Faber theorems

Let the points be assigned on a segment $[a,b]$ that composes an infinite triangle matrix

$$\begin{array}{ccccc} x_1^1 & & & & \\ x_1^2 & x_2^2 & & & \\ x_1^3 & x_2^3 & x_3^3 & & \\ \dots & \dots & \dots & \dots & \\ x_1^n & x_2^n & x_3^n & \dots & x_n^n \\ \dots & \dots & \dots & \dots & \dots \end{array} \tag{5.1}$$

with $x_k^n \neq x_j^n$ for $k \neq j$.

Let us construct for the assigned function $f(x) \in C[a,b]$ a sequence of Lagrange interpolation polynomials $L_{n-1}(x)$ using the points of the nth line in matrix (5.1):

$$L_{n-1}(x) = \sum_{k=1}^{n} f(x_k^n) l_k^n(x), \tag{5.2}$$

where

$$l_k^n(x) = \frac{\omega_n(x)}{\omega_n'(x_k^n)(x - x_k^n)}, \quad \omega_n(x_k^n) = \prod_{k=1}^{n} (x - x_k^n).$$

We regard $L_{n-1}(x)$ as the linear operator transforming the function $f(x) \in C[a,b]$ into an element of the same space: $L_{n-1}f = L_{n-1}(x)$ $(L_{n-1}^2 = L_{n-1})$. Take the value

$$\lambda_n = \max_{x \in [a,b]} \lambda_n(x), \quad \text{where} \quad \lambda_n(x) = \sum_{k=1}^{n} |l_k^n(x)|. \tag{5.3}$$

It easy to show that $\|L_{n-1}\| = \lambda_k$. On the other hand, there is the following important theorem.

Theorem 4 (Bernstein–Faber). *For any matrix* (5.1) *the following inequality is valid:*

$$\lambda_n > \frac{\ln n}{8\sqrt{\pi}}.$$

We accept it, omitting the proof. Therefore, $\|L_{n-1}\| \to \infty$ for $n \to \infty$. Hence, the next theorem is valid too.

Theorem 5 (Faber). *There exists, no matter what the matrix of nodes* (5.1), *a function $f(x) \in C[a,b]$ for which the interpolation polynomial $L_{n-1}(x)$ generated by the nth line of the matrix does not tend uniformly to $f(x)$ for $n \to \infty$.*

Recall that the convergence is understood in the space $C[a,b]$ as a uniform convergence of functions on the segment $[a,b]$. The singularities fixing principle allows for a stronger statement: *there exists $f(x) \in C[a,b]$ for which $\|L_{n-1}f\| \to \infty$, $n \to \infty$ which means that the interpolation method diverges.*

So, the Lagrange polynomial method generates a sequence of projection operators with unbounded norm. However, the following lemma is valid.

Lemma 2. *For $\forall\, f(x) \in C[a,b]$ there exist the nodes* (5.1) *such that $L_{n-1}(x)$ converge to $f(x)$ for $n \to \infty$.*

□ Let $P_{n-1}(x)$ be a polynomial of $(n-1)$th power the least deviating in $C[a,b]$ from $f(x)$. By the alternance theorem, for each $n = 1, 2, \ldots$ there exist n points $x_i^n \in [a,b]$ where $P_{n-1}(x_i^n) = f(x_i^n)$, $i = 1, \ldots, n$. For the table (5.1), formed of these points, $L_{n-1}(x) = P_{n-1}(x)$; consequently, $\|f(x) - L_{n-1}(x)\| = E_{n-1}(f) \to 0$ for $n \to \infty$. ■

Let us estimate an error in $C[a,b]$ for the Lagrange polynomial. We have in terms of Lemma 2,

$$\|L_{n-1}f - f\| = \|L_{n-1}(f - P_{n-1}) + (P_{n-1} - f)\| \le \|f - P_{n-1}\|$$

$$+ \|L_{n-1}(f - P_{n-1})\| \le (1 + \lambda_n)E_{n-1}(f).$$

So, if $\lambda_n E_{n-1}(f) \to 0$ for $n \to \infty$, then $L_{n-1}(f)$ converges to $f(x)$.

Take, for comparison, another approximation method $B_n f = B_n(x)$, where $B_n(x)$ is the Bernstein polynomial (see Example 3.8 in Section 3 of Chapter 1); it provides for uniform approximations for $\forall f(x) \in C[0,1]$; computing $\|B_n\|$, we see that $\|B_n\| = 1$.

3. Application of the Banach–Steinhaus theorem to the representation of functions by integrals and the Fourier–Fejér de la Vallée-Poussin series in the space $C[-\pi,\pi]$

Let a sequence of continuous functions $\{K_n(x,t)\}$ be assigned in a square $[a,b] \times [a,b]$. We say that a function $f(x)$ is *representable by an integral* if a sequence

$$f_n(x) = \int_a^b K_n(x,t) f(t)\,dt, \qquad n = 1, 2, \ldots \tag{5.4}$$

converges in some sense to $f(x)$. Integrals of this kind are met systematically in various aspects of analysis and approximation theory. Take, for example, the integrals of Dirichlet, Fejér, de la Vallée-Poussin, and others; they express finite polynomial and trigonometric sums approximating an assigned function.

Let, for example, $\{\varphi_k\}_1^\infty$ be complete orthonormed with a weight $w(x)$ on an $[a,b]$ system of continuous functions. Then we have for the Fourier polynomials of an order n of a function $f(x)$ (see Section 8 of Chapter 1):

$$S_n(x) = \sum_{k=1}^{n} C_k \varphi_k(x) = \sum_{k=1}^{n} \varphi_k(x) \int_a^b f(t)\varphi_k(t) w(t)\,dt$$

$$= \int_a^b K_n(x,t) f(t)\,dt,$$

where

$$K_n(x,t) = w(t) \sum_{k=1}^{n} \varphi_k(x)\varphi_k(t).$$

Let us consider (5.4) as a sequence of integral operators mapping a space $C[a,b]$ onto $C[a,b]$. We can show that norms of these operators are equal to

$$\max_{a \le x \le b} \int_a^b |K_n(x,t)|\, dt. \tag{5.5}$$

There is the following Banach–Steinhaus theorem.

Theorem 3′. It is necessary and sufficient for a sequence $f_n(x)$ to converge to $f(x)$ in the space $C[a,b]$ for $\forall\, f(x) \in C[a,b]$, that there exists $M > 0$ such that:

(1) $\max\limits_{a \le x \le b} \int_a^b |K_n(x,t)|\, dt \le M$;

(2) $\int_a^b K_n(x,t) f(t) dt \to f(t)$ *for everywhere dense in $C[a,b]$ set of functions (for instance, algebraic or trigonometric polynomials).*

Remark. If $K_n(x,t) \ge 0$ for $a \le x \le b$, then condition (1) of Theorem 3′ is equivalent to the condition

$$\max_{a \le x \le b} \int_a^b K_n(x,t)\, dt \le M. \tag{5.5′}$$

We know that the Fourier series for a function $f(x) \in L_2(-\pi,\pi)$

$$\frac{a_0}{2} + \sum_{k=1}^{\infty} (a_k \cos kx + b_k \sin kx) \tag{5.6}$$

converges in the space $L_2(-\pi,\pi)$ and the values of sums of nth order $S_n(x)$ are the best approximations for $f(x)$. It is known from the analysis course that $S_n(x)$ is expressed in the form of a linear integral operator (the Dirichlet integral)

$$S_n(x) = S_n f = \frac{1}{2\pi} \int_{-\pi}^{\pi} \frac{\sin(2n+1)\frac{t-x}{2}}{\sin\frac{t-x}{2}} f(t)\, dt. \tag{5.7}$$

If $f(x)$ is here a trigonometric polynomial of power n, then $S_n f = f(x)$; that is, the operator $S_n f = S_n(x)$, is an orthoprojector in $L_2(-\pi,\pi)$ and $\|S_n\| = 1$.

Another picture of the series convergence is observed in the space $C[-\pi,\pi]$. In this case condition (2) of Theorem 3′ is satisfied as before, although estimating the operators' norms in (5.7) by formula (5.5) we obtain that

$$\|S_n\| \geq \frac{1}{8\pi}\ln n.$$

We conclude by Theorem 3′, that *there exists a continuous periodic function of period* 2π *whose Fourier series converges uniformly to no function.* The singularities fixing principle allows for a stronger result: *there exists a continuous periodic function* $f(x)$ *whose Fourier series diverges in* $C[-\pi,\pi]$; *that is,* $\|S_n(f)\| \to \infty$, $n \to \infty$.

Take, for comparison, the Fejér summing method [see Example 3.5 (Section 3 of Chapter 1)] producing uniform approximations for
$\forall f(x) \in C[-\pi,\pi]$. For this method

$$\sigma(f,x) = \sigma_n(x) = \frac{1}{n+1}\sum_{i=0}^{n} S_i(x) \tag{5.8}$$

and the function $K_n(x,t)$ in integral representation (5.4) has the form

$$K_n(x,t) = \frac{1}{2(n+1)\pi}\left(\frac{\sin(n+1)\frac{(t-x)}{2}}{\sin\frac{t-x}{2}}\right)^2. \tag{5.9}$$

It is nonnegative and the constant M in inequality (5.5′) equals 1: $\|\sigma_n(f,x)\| \leq \|f\|$.

Take, finally, the third method, the de la Vallée-Poussin method [see Example 3.9 (Section 3 of Chapter 1)]. For this method

$$v_{2n-1}(f,x) = 2\sigma_{2n-1}(f,x) - \sigma_{n-1}(f,x); \tag{5.10}$$

that is,

$$\|v_{2n-1}(f,x)\| \leq 2\|\sigma_{2n-1}(f,x)\| + \|\sigma_{n-1}(f,x)\| \leq 3\|f\| \tag{5.11}$$

and if $P_n(x)$ is any trigonometrical polynomial of nth power, then

$$v_{2n-1}(P_n(x),x) = P_n(x). \tag{5.12}$$

So, it follows from formulas (5.9) and (5.10) that the de la Vallée-Poussin method has an integral representation (5.4), and from (5.11) and (5.12) that conditions of Theorem 3′ are satisfied. Here is the error estimate for this method. Let $\tilde{E}_n(f)$ be the best approximation for $f(x) \in C[-\pi,\pi]$ with trigonometric polynomial $\tilde{P}_n(x)$ of nth power. Then with regard to (5.11) and (5.12) we have

$$|f(x) - v_{2n-1}(f,x)\|$$

$$= \|f - P_n + v_{2n-1}(P_n - f, x)\| \leq \tilde{E}_n(f) + 3\tilde{E}_n(f) = 4\tilde{E}_n(f).$$

The de la Vallée-Poussin method, which is a linear combination of the Fejér methods (5.10), thus hus for $\forall\ f(x) \in C[-\pi,\pi]$ a rather high order of convergence.

We have modeled by examples from items 2 and 3 the most characteristic situation the investigator-calculator encounters while solving one of the main problems in numerical mathematics: Problem V (see Section 3 of Chapter 1) on the choice of the best solution method.

There are two alternative methods. One of them is reliable and provides for finding the approximate value of desired element x that belongs to some a priori defined set C. However, this method has a so-called "saturation" property; it means that the approximations it produces are "uniformly rough"; that is, they do not account any more for individual properties (for instance, such as smoothness if smooth functions are implied) of approximated element $x \in C$. In our case these are the Bernstein and Fejér polynomials whose properties are well investigated in the analysis.

Another method is "capricious"; it cannot approximate all the elements $x \in C$. However, if it is possible to reasonably narrow the class of approximated elements x in the same space [see Example 3.7 of the solution of Problem IV on a class of functions satisfying condition (3.14) in Section 3 of Chapter 1 and the solution of Problem V in item 3, Section 10 of Chapter 1 on the optimization of interpolation nodes location for the Lagrange polynomials], then the estimates (3.18) and (10.44) of Chapter 1 show that this method can provide a new class $C' \in C$ for qualitatively better approximation as compared to the first method.

Our advice is evident: before you employ the methods of the first type for an applied problem, study the properties of the problem under investigation, possess the research algorithms of neighboring divisions of mathematics, and . . . the success (possibly) will come (e.g., in the form of the third method).

§ 6. Linear Functionals and Adjoint Space

Linear functionals in a normed space, boundedness, a norm. Examples. Adjoint space, notion on reflexive space, weak convergence of functionals and elements. Banach–Steinhaus theorem for linear functionals and its application to the question of convergence of quadrature forms. Problem on construction of quadrature formulas ($n \geq 1$), functional of an error. Pólya–Steklov theorems, convergence of Gauss quadrature formulas.

1. Linear functionals in a normed space, the boundedness, a norm. Examples. Adjoint space, notion on reflexive space. Weak convergence of functionals and elements

Let $\mathbf{X}$ be real normed space and let $\mathbf{Y}$ be real axis ($\mathbf{R}^1$). Linear operator $f : \mathbf{X} \to \mathbf{R}^1$ mapping $\mathbf{X}$ onto $\mathbf{R}^1$ is called a *linear functional.* It is denoted by $f(x)$ or $\langle f, x\rangle$ (parenthesis form preserves many features of an inner product). We consider only real functionals. However, there may be functionals defined on complex normed space whose domain of values is all complex plane.

The linearity of a functional means that:

(1) $D(f)$ is linear space;

(2) $f(\alpha x + \beta y) = \alpha f(x) + \beta f(y)$ for $\forall\, x, y \in D(f)$ and $\forall\, \alpha, \beta \in \mathbf{R}^1$. (6.1)

The boundedness means that

$$\|f\| = \sup_{x \in D(f)} \frac{|\langle f, x\rangle|}{\|x\|} < \infty. \tag{6.2}$$

Then

$$|\langle f, x\rangle| \leq \|f\| \cdot \|x\|. \tag{6.3}$$

Formula (6.2) is important since it defines a *norm for elements of adjoint space.*

Examples

6.1. Let $f = (f_1, f_2, \ldots, f_n) \in \mathbf{R}^n$, $x = (x_1, \ldots, x_n) \in \mathbf{R}^n$; then the expression

$$f(x) = (f, x) = \sum_{i=1}^{n} f_i x_i$$

is linear functional in $\mathbf{R}^n$.

6.2. Let $\mathbf{X} = H$ be Hilbert space and let $f \in H$ be some fixed element; then

$$f(x) = (f, x) \tag{6.4}$$

is linear functional in H.

6.3. Let $\mathbf{X} = C[a, b]$, $f(t), x(t) \in C[a, b]$, then the expression $\langle f, x\rangle = \int_a^b f(t)x(t)dt$ or $\langle f, x\rangle = x(t_0)$ where $t_0 \in (a, b)$ are linear functionals; in more general form

$$\langle f, x\rangle = \int_a^b x(t)\, d\sigma(t),$$

where $\sigma(t)$ is a function of bounded value and the integral is the Stieltjes integral.

Let $\mathbf{X}$ be Banach space; take $\mathcal{L}(\mathbf{X}, \mathbf{R}^1)$, the Banach space of linear bounded functionals assigned on $\mathbf{X}$. This space is referred to as *adjoint* and denoted by $\mathbf{X}^*$. By virtue of linearity of $\mathbf{X}$ and $\mathbf{X}^*$ the following equalities are valid for $\forall\, \alpha_1, \alpha_2, \beta_1, \beta_2 \in \mathbf{R}^1$, $\forall\, x_1, x_2$, $x \in \mathbf{X}$, and $\forall\, f_1, f_2, f \in \mathbf{X}^*$:

$$\begin{aligned} \langle \alpha_1 f_1 + \alpha_2 f_2, x\rangle &= \alpha_1\langle f_1, x\rangle + \alpha_2\langle f_2, x\rangle \\ \langle f, \beta_1 x_1 + \beta_2 x_2\rangle &= \beta_1\langle f_1, x_1\rangle + \beta_2\langle f, x_2\rangle. \end{aligned} \tag{6.5}$$

We study, as a rule, only linear functionals; therefore, we sometimes simply call them functionals.

Zero functional $0^* \in \mathbf{X}*$ is defined by the equality

$$\langle 0^*, x\rangle = 0 \quad \text{for} \quad \forall\, x \in \mathbf{X}.$$

Since $\mathbf{X}^*$ is Banach space, linear functionals may be defined in turn on the elements of this space composing thus the second adjoint to $\mathbf{X}$ space $\mathbf{X}^{**}$. If $\mathbf{X}^* = \mathbf{X}$ with accuracy up to the isometry, then $\mathbf{X}^{**} = \mathbf{X}$ too. Now let $\mathbf{X}^* \neq \mathbf{X}$. Let us find out what elements may compose the space $\mathbf{X}^{**}$; to this end take the functionals

$$\langle f, x\rangle \tag{6.6}$$

where we fix an element $x \in \mathbf{X}$ and change $f \in \mathbf{X}^*$. Then by the first property (6.5) it will be linear functional defined on the elements $\mathbf{X}^*$; its norm according to (6.3) does not exceed (as a matter of fact, it equals $\|x\|$). There is one-to-one correspondence between each linear functional (6.6) and an element $x \in \mathbf{X}$ preserving the correspondence of results of summing and multiplication by a scalar operation [see the second relationships in (6.5)]. Consequently, the space $\mathbf{X}$ is isomorphic and isometric to some subspace in $\mathbf{X}' \subset \mathbf{X}^{**}$; that is, $\mathbf{X} \subset \mathbf{X}^{**}$ with accuracy up to isomorphism and isometry.

If $\mathbf{X}^{**} = \mathbf{X}$, then the Banach space $\mathbf{X}$ is termed *reflexive*. Such spaces possess many good features of Hilbert spaces therefore ply an important role in applications.

It is possible to define in the spaces $\mathbf{X}$ and $\mathbf{X}^*$ a new type of convergence, namely, *weak convergence*. We say that a sequence $\{x_n\} \in \mathbf{X}$ converges to $x \in \mathbf{X}$ *weakly* in $\mathbf{X}$ ($x_n \mapsto x$) if for $\forall\, f \in \mathbf{X}^*$ $\langle f, x_n \rangle \to \langle f, x \rangle$ for $n \to \infty$. We say that the sequence $\{f_n\} \in \mathbf{X}^*$ converges to $f \in \mathbf{X}^*$ *weakly in* $\mathbf{X}^*$ ($f_n \mapsto f$) if for $\forall\, x \in \mathbf{X}$ $\langle f_n, x \rangle \to \langle f, x \rangle$ for $n \to \infty$.

We had defined before the notion of strong convergence to $A \in \mathcal{L}(\mathbf{X}, \mathbf{Y})$ of a sequence of operators $\{A_n\} \in \mathcal{L}(\mathbf{X}, \mathbf{Y})$ with the condition that $\forall\, x \in \mathbf{X}$ $\|A_n x - Ax\| \to 0$ for $n \to \infty$. Let us make clear what does this notion would mean as related to a sequence of linear functionals $\{f_n\} \in \mathcal{L}(\mathbf{X}, \mathbf{R}^1)$

$$\|\langle f_n, x \rangle - \langle f, x \rangle\| = |\langle f_n, x \rangle - \langle f, x \rangle| \to 0$$

for $n \to \infty$ for $\forall\, x \in \mathbf{X}$. We see that the notion of strong convergence applied to linear functionals is *equivalent* (is a synonym) to the notion of weak convergence of functionals. Let us formulate, according to this remark, the Banach–Steinhaus theorem especially for linear functionals.

The elements $x \in \mathbf{X}$, $f \in \mathbf{X}^*$ are referred to as *biorthogonal* if $\langle f, x \rangle = 0$. The sequences $\{x_n\}$ from $\mathbf{X}$ and $\{f_n\}$ from $\mathbf{X}^*$ are termed *biorthogonal* if $\langle f_i, x_j \rangle = \delta_{ij}$. Let the elements $x \in \mathbf{X}$ and $f \in \mathbf{X}^*$ be representable on such sequences in the form of the series

$$x = \sum_{i=1}^{\infty} a_i x_i, \qquad f = \sum_{i=1}^{\infty} b_i f_i.$$

It is easily seen that $\langle f_i, x \rangle = a_i$ and $\langle f, x_i \rangle = b_i$; that is,

$$x = \sum_{i=1}^{\infty} \langle f_i, x \rangle, \qquad f = \sum_{i=1}^{\infty} \langle f, x_i \rangle f_i.$$

These series are called the *Fourier series* on biorthogonal sequences $\{x_n\}$, $\{f_n\}$. The first interesting applications of biorthogonal sequences to interpolation problems were proposed by Chebyshev and Markov.

2. The Banach–Steinhaus theorem for linear functionals and its application to the question of the convergence of quadrature forms. Problem on the construction of quadrature formulas, the theorems by Pólya and Steklov, the convergence of Gauss quadrature formulas

Theorem 1 (by Banach–Steinhaus for linear functionals). Let $\mathbf{X}$ *be a Banach space and let* $\{f_n\}_1^\infty$ *and* f *be linear functionals from* $\mathbf{X}^*$. *It is necessary and sufficient for* $f_n \overset{\text{weak}}{\mapsto} f$ *that:*

(1) *norms* $\|f_n\|$ *should be bounded as an aggregate;*

(2) $\langle f_n, x\rangle \to \langle f, x\rangle$ *for all* x *from some set* $\mathbf{X}'$ *everywhere dense in* $\mathbf{X}$.

They use, for the approximate calculation of integrals

$$I(f) = \int_a^b f(x)w(x)\,dx, \qquad f(x) \in C[a,b], \tag{6.7}$$

where $w(x)$ is an integrable positive on an (a,b) weight function, the so-called *quadrature formulas*

$$I_N(f) = \sum_{k=0}^{N} A_k^N f(x_k^N), \tag{6.8}$$

where the points x_k^N belong, as a rule, to the segment $[a,b]$ and are called the *nodes* ($x_k^N \neq x_i^N$ for $k \neq i$) and the numbers A_k^N are called the *weights* of the quadrature formula (6.8). For example, such are the rectangles, trapeziums, Simpson, Newton–Cotes, and Gauss formulas and the like.

If for $\forall f \in C[a,b]$ $I_N(f) \to I(f)$, $N \to \infty$, then we say that the *quadrature process converges*. The values $I(f), I_N(f)$ are linear functionals defined on functions of the space $C[a,b]$. Let us estimate norms of the functionals $I_N(f)$ in this space:

$$|I_N(f)| \le \sum_{k=0}^{N} |A_k^N|\,|f(x_k^N)| \le \sum_{k=1}^{N} |A_k^N| \max_k |f(x_k^N)|; \tag{6.9}$$

that is,

$$\|I_N\| \le \sum_{k=1}^{N} |A_k^N|.$$

We have, as a matter of fact, exact equality

$$\|I_N\| = \sum_{k=1}^{N} |A_k^N|. \tag{6.10}$$

In order to check it, one should take in (6.9) continuous piecewise linear function $f(x)$ such that $f(x_k^N) = \text{sign}\, A_k^N$, $f(a) = f(b) = 1$ if the points $x = a$, $x = b$ do not belong to the number of nodes of the quadrature formula. For this function, its diagram is a continuous broken line and the system of inequalities (6.9) turns into a system of equalities.

There exist various definition principles for the nodes and weights in quadrature formula (6.8). Take one of them. Let $\{u_i\}_0^\infty$ be complete in a $C[a,b]$ system of continuous linearly independent functions and let it be such that for $\forall\, N \ge 0$ the functions $\{u_i\}_0^N$ satisfy on $[a,b]$ the Chebyshev conditions (see Section 9 of Chapter 1) and it is possible to calculate integrals (6.7) for $f(x) = u_i(x)$ exactly. Let $A_N = \{a_{ij}\}$ be a matrix of an order N with elements $a_{ij} = u_i(x_j^N)$. It is easy to show for such systems that: (a) $\det A \ne 0$ for different x_j^N and (b) function $u_0(x) \ne 0$ on $[a,b]$; therefore, to be definite, we regard it as positive and assume that $u_0(x) \ge c_0 > 0$. We call a generalized polynomial of kind

$$P_m(x) = \sum_{i=0}^{m} a_i u_i(x), \qquad a_m \ne 0 \tag{6.11}$$

the polynomial of mth order. The system of powers $u_i(x) = x^i$ for $w(x) = 1$ can serve as an example of such systems. We assume here that the nodes and weights of a quadrature formula are determined from the condition under which the quadrature formula is exact only for any polynomials of an order not higher than $m = \varphi(N)$ where $\varphi(N)$ is some integer positive function tending to infinity for $N \to \infty$; that is,

$$I_N(P_m) = I(P_m) \tag{6.12}$$

for all polynomials $P_m(x)$ of power not higher than m. The number m is called the *algebraic degree of accuracy* of the quadrature formula. Substituting $f(x) = u_i(x)$, $i = \overline{0,m}$ into equality (6.12) we obtain $(m+1)$ of equations linking parameters A_k^N, x_k^N, $k = \overline{1,N}$ of the quadrature formula:

$$\sum_{k=0}^{N} A_k^N u_i(x_k^N) = I(u_i), \qquad i = 0, 1, \ldots, m. \tag{6.13}$$

Theorem 2 (the Pólya theorem). *Let for each N quadrature formula* (6.8) *be exact for all polynomials of power not higher than $m = \varphi(N)$, $\varphi(N) \to \infty$ for $N \to \infty$. Then it is necessary and sufficient for a quadrature process to converge for any function $f(x) \in C[a,b]$, that there exists a constant M such that*

$$\sum_{k=0}^{N} |A_k^N| \leq M \quad for \quad \forall\, N \geq 0. \tag{6.14}$$

□ By (6.10) and (6.14) $\{\|I_N\|\}$ is bounded, and by virtue of (6.12) we conclude that $I_N(P) \to I(P)$ for $N \to \infty$ for any polynomial from an everywhere dense in $C[a,b]$ set of all polynomials; therefore, the theorem is valid due to the Banach–Steinhaus theorem for functionals. ■

There exists a class of quadrature formulas (6.8) with nonnegative weights: $A_k^N \geq 0$; the following theorem is valid in this case.

Theorem 3 (the Steklov theorem). *Let for each N formula,* (6.8) *be exact for all polynomials of power not higher than $m = \varphi(N)$, $\varphi(N) \to \infty$ for $N \to \infty$ and let all its coefficients $A_k^N \geq 0$. Then the quadrature process converges for any continuous function.*

□ Let us check if the conditions of Theorem 2 are satisfied. If $f(x) = u_0(x)$, then equality (6.13) for $i = 0$ must be valid, which determines a value of M in (6.14):

$$M \leq c_0^{-1} \int_a^b u_0(x) w(x)\, dx. \blacksquare$$

The system of equations (6.13) defines quadrature forms of various types. If $m = N - 1$ there and the nodes $\{x_k^N\}$ are assumed to be assigned, then we get formulas of the Newton–Cotes type. If n_1 of nodes are known and n_2 of nodes are to be determined ($n_1 + n_2 = N$) for $m = N + n_2 - 1$, then (6.13) is a nonlinear system of equations, and if it is resolvable, we obtain the Lobatto–Markov type formulas. Regarding all parameters $\{A_k^N, x_k^N\}_1^N$ as to be determined and assigning $m = 2N - 1$ in (6.13) we arrive at a nonlinear system of equations; it is known from the theory of generalized momentums, that there exists a solution to this system with $A_k^N > 0$, and the nodes $\{x_k^N\}_1^N \in (a, b)$ and realize so-called *lower main representation*. Then quadrature formula (6.8) is called a *quadrature formula of Gauss type*. In particular, for $u_i(x) = x^i$ for $a = -1$, $b = 1$, $w(x) = 1$ the nodes are the roots of the Legendre polynomial of Nth power, and for $w(x) = (1-x^2)^{-1/2}$ we obtain the remarkable Mehler formula with equal weights and nodes in the roots of the Chebyshev polynomial:

$$A_k^N = \frac{\pi}{N}, \qquad x_k^N = \cos\frac{2k-1}{2N}.$$

Remark 1. For the so-called composite quadrature rectangle, trapezium, and Simpson formulas we can take as everywhere dense in $C[a, b]$ sets the sets of piecewise constant, piecewise linear, and, finally, piecewise quadrature functions and justify, in the same way, the convergence of these classes of quadrature formulas that are exact on corresponding everywhere dense sets.

Remark 2. Peculiarities of a class of functions to be integrated can be taken into account by the choice of a weight function and the system $\{u_i\}_0^\infty$.

§ 7. The Riesz Theorem. The Hahn–Banach Theorem. Optimization Problem for Quadrature Formulas. The Duality Principle

Riesz theorem on general form of linear functionals and Hahn–Banach theorem on the continuation of linear functional for Hilbert spaces. Application of Riesz theorem to the optimization problem for quadrature formulas. The duality principle for extremal problems.

The Riesz theorem and the Hahn–Banach theorem are the main theorems in functional analysis; they are widely used in various divisions of mathematics. We present the proofs only for the case of Hilbert spaces, although the theorems are valid for spaces of rather general form.

1. The Riesz theorem on general form of linear functionals in Hilbert space

Theorem 1 (*the Riesz theorem*). *Let H be Hilbert space. For any linear bounded functional f assigned everywhere on H there exists a unique element $y \in H$ such that for $\forall\ x \in H$*

$$\langle f, x\rangle = (y, x) \tag{7.1}$$

with $\|f\| = \|y\|$.

□ Let $L = \ker f$; that is, a set of all elements $z \in H$ such that $\langle f, z\rangle = 0$. The set L consists either of one zero element or is a subspace H. If $L = H$, then $\|f\| = 0$ and it is possible to take $y = 0$.

Let $L \neq H$; then at least one element $v \in H$ will be found such that $\langle f, v\rangle \neq 0$ and the expansion $v = z + q$ will be valid where $z \in L$, $q \in L^{\perp}$; that is, $\langle f, v\rangle = \langle f, q\rangle$; therefore, $\langle f, q\rangle \neq 0$. For $\forall x \in H$ we have

$$\langle f, x - \frac{\langle f, x\rangle}{\langle f, q\rangle} q\rangle = 0;$$

that is, the element

$$x - \frac{\langle f, x\rangle}{\langle f, q\rangle} q \in L$$

and therefore it is orthogonal to q and, consequently,

$$(q, x) = \frac{\langle f, x\rangle}{\langle f, q\rangle}(q, q)$$

hence

$$\langle f, x\rangle = \frac{\langle f, q\rangle}{(q, q)}(q, x) = (y, x),$$

where $(\langle f, q\rangle / \langle q, q\rangle) q$. Let us show now that $\|f\| = \|y\|$. In fact,

$$|\langle f, x\rangle| = |(y, x)| \leq \|y\|\,\|x\|$$

therefore, by definition of the f norm we have $\|f\| \leq \|y\|$ and, besides,

$$\langle f, y\rangle = (y, y) = \|y\|^2 \leq \|f\|\,\|y\|$$

whence $\|y\| \leq \|f\|$. Therefore, $\|f\| = \|y\|$.

Let there exist $\bar{y} \in H$, $\bar{y} \neq y$ such that $\langle f, x \rangle = (\bar{y}, x)$. Then $(y - \bar{y}, x) = 0$ for $\forall\, x \in H$; that is, $y = \bar{y}$. The contradiction so obtained proves the uniqueness of the representation (7.1). ■

Corollary 1. *Hilbert space H is selfadjoint space; that is, $H^* = H$.*

□ The Riesz theorem establishes one-to-one correspondence between the spaces H and H^* preserving a norm and results of linear operations: if $f_1 \leftrightarrow y_1$, $f_2 \leftrightarrow y_2$, then $\lambda_1 f_1 + \lambda_2 f_2 \leftrightarrow \lambda_1 y_1 + \lambda_2 y_2$; this means that the spaces H and H^* coincide with accuracy up to isometric isomorphism. ■

2. The Hahn–Banach theorem on the continuation of the linear functional with its norm being preserved

Theorem 2 (*the Hahn–Banach theorem*). *Let L be a closed subspace of a Banach space $\mathbf{X}$. If f is a linear continuous functional defined on L, then there exists a continuous linear functional g defined on $\mathbf{X}$ which is a continuation of f with the same norm that f has:*

$$\langle g, x \rangle = \langle f, x \rangle \quad for \quad \forall\, x \in L, \qquad \|g\|_{\mathbf{X}} = \|f\|_L. \tag{7.2}$$

□ Prove, for the sake of simplicity, the theorem when $\mathbf{X}$ is the Hilbert space H. Let P be an orthoprojector from H onto L. Take, for $\forall\, x \in H$ a functional

$$\langle g, x \rangle = \langle f, Px \rangle. \tag{7.3}$$

It satisfies the conditions (7.2) since firstly, $\langle g, x \rangle = \langle f, x \rangle$ for $x \in L$, as $Px = x$, and secondly, two inequalities hold:

$$\|f\|_L = \sup_{v \in L} \frac{|\langle g, v \rangle|}{\|v\|} \leq \sup_{v \in H} \frac{|\langle g, v \rangle|}{\|v\|} = \|g\|_H$$

and, since $\|Pv\| \leq \|v\|$ for $\forall\, v \in H$, then, with regard to (7.3), we have

$$\|g\|_H = \sup_{v \in H} \frac{|\langle g, v \rangle|}{\|v\|} \leq \sup_{Pv \in L} \frac{|\langle f, Pv \rangle|}{\|Pv\|} = \|f\|_L$$

from which it follows that $\|g\|_H = \|f\|_L$. ■

Corollary 2. *For $\forall\, x_0 \neq 0$ from $\mathbf{X}$ there exists $f_0 \in \mathbf{X}^*$ such that $\|f_0\| = 1$ and $\langle f_0, x_0 \rangle = \|x_0\|$.*

□ Let L be a subspace of elements $x = tx_0$ for all real t. Set $\langle f_0, x\rangle = t\|x_0\|$ on L. Now, extend this functional to all $\mathbf{X}$ preserving the unit norm. ■

3. The problem of the construction of quadrature formulas: the functional statement of the problem on optimization of quadrature formulas

Take once more the problem of the construction of quadrature formulas for approximate calculation of integrals that are multidimensional this time. Let Ω be a bounded domain in $\mathbf{R}^n$ with a sufficiently smooth boundary and let $x = (x_1, \ldots, x_n)$,

$$I(f) = \int\limits_{\Omega} f(x)dx \tag{7.4}$$

$$I_N(f) = \sum_{k=1}^{N} A_k^N f(x_k^N) \tag{7.5}$$

be the *quadrature* (*or cubature*) *formula* for the calculation of integral (7.4) where the points $x_k^N = (x_{1k}^N, x_{2k}^N, \ldots, x_{nk}^N)$, that are the nodes of quadrature formula (7.5), belong, as a rule, to the domain Ω and $N \geq 1$.

Let $\chi\Omega(x)$ be the characteristic function of the domain $\Omega : \chi_\Omega(x) = 1$ if $x \in \bar{\Omega}$ and $\chi_\Omega(x) = 0$ if $x \neq \bar{\Omega}$; then

$$I(f) = \int\limits_{\mathbf{R}^n} f(x)\chi_\Omega(x)\, dx. \tag{7.6}$$

The difference

$$\langle l_N, f\rangle = \int\limits_{\mathbf{R}^n} f(x)\chi_\Omega(x)dx - \sum_{k=1}^{N} A_k^N f(x_k^N) \tag{7.7}$$

is called an *error of quadrature formula*; it is linear functional in the space $C(\bar{\Omega})$. We say that the quadrature formula is exact on a function $f \in C(\bar{\Omega})$ if $\langle l_N, f\rangle = 0$; the functionals, upon which the quadrature formula is exact, thus compose a kernel of a functional of an error ($\ker l_N$).

The algebraic statement of the problem on the construction of the quadrature formula is as follows. The weights A_k^N, $k = \overline{1, N}$ are to be found for a given system of nodes x_k^N, $k = \overline{1,N}$ (or the weights A_k^N and x_k^N, $k = \overline{1,N}$) in such a way that $\langle l_N, x^\alpha \rangle = 0$, $1 \leq |\alpha| \leq m$ where $x^\alpha = x_1^{\alpha_1} x_2^{\alpha_2} \dots x_n^{\alpha_n}$, $|\alpha| = \sum_{i=1}^{n} \alpha_i$ for an as large as possible value of m. Let L_m be the linear hull of such x^α; $L_m \subset \ker l_N$. We met such a statement in the preceding section; therefore, the definition of convergence and Theorems 2 and 3 of Section 6 stay valid (due to the completeness of x^α in $C(\bar{\Omega})$) for the multidimensional case.

Now take the functional statement of the problem on optimization of a quadrature formula posed by S. Sobolev. Let H be a Hilbert space with an inner product $(\cdot, \cdot)$, consisting of some elements of the space $C(\bar{\Omega})$. Let us assume that in (7.4) and (7.5) $f \in H \subset C(\bar{\Omega})$ [the space H can be, for example, the Sobolev space $W_2^l(\Omega)$ for $2l > n$; we investigate some properties of these spaces in Section 14]. The problem on the optimization of quadrature formula (7.5) on a class of functions $f(x) \in H$ means finding a functional $\langle l_N, f \rangle$ for given N such that its norm in the space H^* is minimal. The choice of a space H permits some arbitrariness; it is determined by additional considerations related to the class of integrable functions. Let

$$w_N = \inf \|l_N\|, \tag{7.8}$$

where inf is taken over $\{A_k^N\}$ or over $\{A_k^N\}$ and $\{x_k^N\}$. This is optimal on the H estimate of an error of quadrature formula. If there exists a quadrature formula for which $\|l_N\| = w_N$, then such a quadrature formula is referred to as *optimal or the best on a class of functions from H*. A function $\varphi(x) \in H$ is called the *extremal function* of a functional $\langle l_N, f \rangle$ if $|\langle l_N, \varphi \rangle| = \|l_N\| \cdot \|\varphi\|_H$.

The Riesz theorem helps to express an extremal function with assigned functional l_N since it can be represented in the form $\langle l_N, f \rangle = (h_N, f)$ where $h_N \in H$; then $\|l_N\| = \|h_N\|$ and $h_N \perp \ker l_N$. Notice that $h_N = h_N(\{A_k^N\}, \{x_k^N\})$. The extremal function φ is determined as follows: $\varphi = h_N$ since in the general case $|(h_N, f)| \leq \|h_N\| \cdot \|f\|$ and for $f = h_N$ this inequality turns into the equality

$$(h_N, h_N) = \|h_N\|^2.$$

If a unit ball $\overline{D}_1(0) \subset H$ is compact in $C(\overline{\Omega})$ and the problem on the construction of quadrature formulas is resolvable in an algebraic statement for any m with $A_k^N \geq 0$, then it is possible, using Theorems 1 and 2 from Section 13 of Chapter 2, to find the upper bound $\|h_N\|$ for $m \to \infty$ and the lower bound $\|h_N\|$ is expressed through the Kolmogorov N-width $D_N(\bar{D}_1(0), C(\bar{\Omega}))$.

4. The duality principle

While investigating some extremal problems, it is sometimes possible to connect with them certain dual problems that facilitate the solution of initial problems. Due to this principle, a unit approach was developed to a rather wide class of problems in the theory of best approximations, optimal control, the estimation of functionals, and so forth.

Let n linearly independent elements $\{u_i\}_1^n$ be assigned in a normed space $\mathbf{X}$. Then for any system of numbers $\{c_i\}_1^n$ with $\sum_{i=1}^{n} |c_i| > 0$ the two following problems can be taken.

Problem A: Find

$$\frac{1}{M} = \min \|\sum_{i=1}^{n} \xi_i u_i\| \tag{7.9}$$

under the condition

$$\sum_{i=1}^{n} \xi_i c_i = 1. \tag{7.10}$$

*Problem A**: Find a minimum of norms of linear bounded functionals $f(x)$ satisfying the relationships

$$f(u_k) = c_k, \qquad k = 1, \ldots, n; \tag{7.11}$$

that is, find

$$m = \inf_{f(u_k)=c_k} \|f\|. \tag{7.12}$$

It is easily verified that a minimum is attainable in Problem A.

Theorem 3 (the duality principle). A minimum in Problem A coincides with a value inverse to a minimum in Problem A; that is, $m = M$.*

□ Let the functional $f(x)$ satisfy (7.11); then for $\forall\, \xi_i,\ i = 1, \ldots, n$

$$|\sum_{i=1}^{n} \xi_i c_i| = |f(\sum_{i=1}^{n} \xi_i u_i)| \leq \|f\| \, \|\sum_{i=1}^{n} \xi_i u_i\|$$

whence

$$\frac{1}{\|f\|} \leq \min_{\xi_i} \frac{\|\sum_{i=1}^{n} \xi_i u_i\|}{|\sum_{i=1}^{n} \xi_i c_i|} = \min_{\sum_{i=1}^{n} \xi_i c_i = 1} \|\sum_{i=1}^{n} \xi_i u_i\| = \frac{1}{M};$$

that is, $\|f\| \geq M$ and, consequently, $m \geq M$.

Now, let us define the functional $f(x)$ in linear n-dimensional space $L_n \subset \mathbf{X}$ spanned over the elements $u_1, u_2, \ldots, u_n$ with the formula $f(\sum_{i=1}^{n} \xi_i u_i) = \sum_{i=1}^{n} \xi_i c_i$. Then a norm of $f(x)$ in L_n looks like

$$\|f\|_{L_n} = \sup_{\xi_i} \frac{|\sum_{i=1}^{n} \xi_i c_i|}{\|\sum_{i=1}^{n} \xi_i u_i\|} = M.$$

Extend, by the Hahn–Banach theorem, the functional $f(x)$ to all space $\mathbf{X}$ preserving a norm. Conditions (7.11) are satisfied; therefore, $m \leq M$. We obtain as a result, that $m = M$. ■

Denote by $l = l(c)$ the common value of m, M, determined by the sequence $c = \{c_k\}_1^n$, and call the *problem* L *in the space* $\mathbf{X}$ the following problem.

Problem L. Let c and L be given. Find under what conditions there exist linear functionals $f(x)$ satisfying the relationships

$$f(u_k) = c_k, \quad k = 1, 2, \ldots, n, \qquad \|f\| \leq L. \tag{7.13}$$

By Theorem 3, the following corollary gives an answer.

Corollary 2. Linear functional $f(x)$ *satisfying conditions* (7.13) *exists, iff* $L \geq l(c)$.

Remember, that u is an extremal element of the functional $f(x)$ if

$$|f(u)| = \|f\| \, \|u\|$$

is valid for it.

Theorem 4. *It is necessary and sufficient for an element* $u = \sum_{i=1}^{n} \xi_i u_i$ *with* $\sum_{i=1}^{n} \xi_i c_i = 1$ *to be a minimizing element for Problem A, that the element u would be the extremal element of some arbitrarily chosen minimal with respect to the norm solution of Problem L (7.3):*

$$f(u_k) = c_k, \quad k = 1, \ldots, n, \qquad \|f\| = l(c). \tag{7.14}$$

□ Let $f_0(x)$ be some solution of (7.14) and let $v = \sum_{i=1}^{n} \alpha_i u_i$ be some minimizing element of Problem A; that is, $\sum_{i=1}^{n} \alpha_i c_i = 1$ and $\|v\| = 1/l(c)$; then

$$1 = \sum_{i=1}^{n} \alpha_i c_i = f_0(v) = \|f_0\| \, \|v\| \tag{7.15}$$

since $\|f_0\| = l(c)$.

Inversely, if $v = \sum_{i=1}^{n} \alpha_i u_i$ with $\sum_{i=1}^{n} \alpha_i c_i = 1$ is the extremal element of the functional $f_0(x)$, then equality (7.15) is valid and, consequently, $\|v\| = 1/l(c)$. ■

§ 8. Adjoint, Selfadjoint, Symmetric Operators

Adjoint operator, linearity, boundedness, norm. Adjoint operator in Hilbert space. Selfadjoint operator, properties of a sum and product, power and polynomial of selfadjoint operator. Symmetric operator. Example. Nonnegative and positive definite operators. Generalized Cauchy–Bunyakovskii inequality. Properties of an operator of orthogonal projecting. A norm of selfadjoint operator, lower and upper bounds of this operator.

1. Adjoint operator, linearity, boundedness, norm. Adjoint operator in Hilbert space. Examples

Let $A \in \mathcal{L}(\mathbf{X}, \mathbf{Y})$ and let g be a continuous linear functional in the space $\mathbf{Y}$, that is, an element of adjoint space $\mathbf{Y}^*$. Assign for arbitrary $x \in \mathbf{X}$

$$\langle f, x \rangle = \langle g, Ax \rangle \tag{8.1}$$

$$Ax = y. \tag{8.2}$$

Expression (8.1) is a continuous linear functional with

$$\|f\| \le \|A\| \cdot \|g\|. \tag{8.3}$$

Formula (8.1) associates each functional $g \in \mathbf{Y}^*$ to some functional $f \in \mathbf{X}^*$. An operator performing this association is termed *adjoint* as related to given operator A and denoted by A^*; that is,

$$A^*g = f \tag{8.4}$$

means (8.1) or, in other words, $A^*g = gA$. Equation (8.4) is referred to as *adjoint* to Equation (8.2). So, each $A \in \mathcal{L}(\mathbf{X}, \mathbf{Y})$ has corresponding adjoint operator A^* mapping $\mathbf{Y}*$ into $\mathbf{X}^*$ as shown in the following diagram.

$$\begin{array}{ccc} x \in \mathbf{X} & \overset{\overset{y=Ax}{\longrightarrow}}{\underset{A}{\overline{}}} & y \in \mathbf{Y} \\ \vdots & & \vdots \\ \vdots & & \vdots \\ f = \langle f, x\rangle \in \mathbf{X}^* & \overset{A^*}{\underset{f = A^*g}{\overline{\longleftarrow}}} & g = \langle g, y\rangle \in \mathbf{Y}^* \end{array}$$

$$\Updownarrow$$

$$f = \langle f, x\rangle = \langle A^*g, x\rangle = \langle g, Ax\rangle.$$

Let us show that $A^* \in \mathcal{L}(\mathbf{Y}^*, \mathbf{X}^*)$.

Theorem 1. *Adjoint operator A^* is a bounded linear operator mapping the space $\mathbf{Y}^*$ into the space $\mathbf{X}^*$ and $\|A^*\| = \|A\|$.*

□ Let us check the linearity of A^*. If $g = \lambda_1 g_1 + \lambda_2 g_2$, $g_1, g_2 \in \mathbf{Y}^*$ and $f = A^*g$, then $D(A^*) = \mathbf{Y}^*$ and $f(x) = \langle f, x\rangle = \langle A^*g, x\rangle = \langle g, Ax\rangle = \lambda_1\langle g_1, Ax\rangle + \lambda_2\langle g_2, Ax\rangle = \lambda_1\langle A^*g_1, x\rangle + \lambda_2\langle A^*g_2, x\rangle$; that is,

$$A^*g = \lambda_1 A^*g_1 + \lambda_2 A^*g_2.$$

The boundedness of A^* follows from inequality (8.3) due to which $\|A^*\| \le \|A\|$.

Let $x_0 \in \mathbf{X}$ and $Ax_0 \ne 0$. By Corollary 2 of Section 7 there exists $f_0 \in \mathbf{X}^*$ such that $\|f_0\| = 1$ and $\langle f_0, Ax_0\rangle = \|Ax_0\|$. Then

$\|Ax_0\| = \langle f_0, Ax_0\rangle = \langle A^*f_0, x_0\rangle \leq \|A^*f_0\|\,\|x_0\| \leq \|A^*\|\,\|f_0\|\,\|x_0\| = \|A^*\|\,\|x_0\|$; that is, $\|A\| \leq \|A^*\|$. ■

Let $\{f_k\}_1^n \in \mathbf{X}^*$ and $\{x_k\}_1^n \in \mathbf{X}$ be two linearly independent systems of elements in H. A linear operator of type

$$\bar{P}_n x = \sum_{k=1}^{n} \langle f_k, x\rangle x_k \tag{8.5}$$

is referred to as *finite-dimensional* (and *n-dimensional*). An operator adjoint to $\bar{P}_n$ can be found with the formula

$$\langle f, \bar{P}_n x\rangle = \sum_{k=1}^{n} \langle f_k, x\rangle\langle f, x_k\rangle = \langle \sum_{k=1}^{n} \langle f, x_k\rangle f_k, x\rangle;$$

that is,

$$\bar{P}_n^* y = \sum_{k=1}^{n} \langle f, x_k\rangle f_k \tag{8.5$'$}$$

which means that $\bar{P}_n^*$ is a finite-dimensional operator also.

Let $\mathbf{X} = \mathbf{Y} = H$; then $\mathbf{X}^* = \mathbf{Y}^* = H$ too, and $A \in \mathcal{L}(H)$. Then, by the Riesz theorem, the functionals $\langle f, x\rangle$, $\langle g, y\rangle$ can be identified with the elements $f, g \in H$ such that $\langle f, x\rangle = (f, x)$ and $\langle g, y\rangle = (g, y)$; therefore, equality (8.1), defining an adjoint operator, takes the form

$$(Ax, g) = (x, A^*g) \tag{8.6}$$

for $\forall g \in H$. This equality can be taken as a *definition of adjoint operator* (whose existence we have proven) in Hilbert space: the operator $A^* \in \mathcal{L}(H)$ is referred to as *adjoint* to the operator $A \in \mathcal{L}(H)$ in Hilbert space H if for any elements $x, g \in H$ equality (8.5) is satisfied.

It is possible to show that $(A+B)^* = A^* + B^*$, $(AB)^* = B^*A^*$,

$(A^{-1})^* = (A^*)^{-1}$ if A^{-1} does exist.

Examples

8.1. $\mathbf{X} = \mathbf{Y} = H = \mathbf{R}^n$, $A = \{a_{ij}\}$ is a matrix with real coefficients; then $A^* = \{a_{ji}\}$ is the transposed matrix.

8.2. $\mathbf{X} = \mathbf{Y} = H = L_2(a,b)$, $Ax(t) = \int_a^b K(t,s)x(s)ds$ is the integral operator with real continuous on the square $\{a \le t \le b, a \le s \le b\}$ kernel $K(t,s)$; then for $x(t), g(t) \in L_2(a,b)$ we have, by the Fubini theorem on the change of integration order with respect to t and s:

$$(Ax,g) = \int_a^b \int_a^b K(t,s)x(s)ds \cdot g(t)dt$$

$$= \int_a^b x(s) \int_a^b K(t,s)g(t)dtds = (x, A^*g),$$

where the operator A^* is determined by the formula

$$A^*g(s) = \int_a^b K(t,s)g(t)dt.$$

Thus the integral operator with kernel $K^*(t,s) = K(s,t)$ is an adjoint one to the integral operator with kernel $K(t,s)$.

The following series of problems is often posed: determine for a given value of $f \in \mathbf{X}^*$ a sequence of functionals $\langle f, x_i \rangle$, $i = 1, \dots, n$, where x_i are solutions of Equation (8.2) for $y = y_i$, $i = 1, \dots, n$.

The direct method for the solution of such problems for large n is rather burdensome. We present a method developed by G. Marchuk using the notion of an adjoint equation: first, we solve once for an assigned f adjoint Equation (8.4) and, having found $g \in \mathbf{Y}^*$, calculate desired functionals by formula (8.1): $\langle f, x_i \rangle = \langle g, y_i \rangle$, $i = 1, \dots, n$. In a series of applications the element g is called a *value function* since it permits the estimation of a change $\langle f, x_i \rangle$ under variations y_i.

2. Selfadjoint operator, properties of a sum and product, power and polynomial of selfadjoint operator. Symmetric operator. Example. Nonnegative and positive definite operators. Generalized Cauchy–Bunyakovskii inequality. Properties of an operator of orthogonal projection

An operator $A \in \mathcal{L}(H)$ is termed *selfadjoint* (*or Hermitian*) if $A = A^*$; that is, if for $\forall\, x, y \in H$ the following equality holds:

$$(Ax, y) = (x, Ay). \tag{8.7}$$

Lemma 1. *If A and B are selfadjoint operators, then for any scalars λ_1, λ_2 the operator $\lambda_1 A + \lambda_2 B$ is selfadjoint.*

□ In fact,

$$((\lambda_1 A + \lambda_2 B)x, y) = \lambda_1(Ax, y) + \lambda_2(Bx, y)$$

$$= \lambda_1(x, Ay) + \lambda_2(x, By) = (x, (\lambda_1 A + \lambda_2 B)y). \blacksquare$$

Notice that if $A \in \mathcal{L}(H)$, then the operators A^*A and AA^* are selfadjoint. Indeed, for $\forall\, x, y \in H$,

$$(A^*Ax, y) = (Ax, Ay) = (x, A^*Ay)$$

and

$$(AA^*x, y) = (A^*x, A^*y) = (x, AA^*y).$$

Theorem 2. *If A and B are selfadjoint operators, then the operator AB is selfadjoint iff the operators A and B commute; that is, $AB = BA$.*

□ If A, B are selfadjoint operators, then for $\forall\, x, y \in H$ we have

$$(ABx, y) = (Bx, Ay) = (x, BAy) = (x, (AB)^*y).$$

Therefore, if AB is selfadjoint, then $AB = BA$, and if A and B commute, then $(AB)^* = BA = AB$, that is, AB is a selfadjoint operator. ■

Two corollaries follow from Lemma 1 and Theorem 2.

Corollary 1. *If $A \in \mathcal{L}(H)$ is a selfadjoint operator, then A^n is a selfadjoint operator for any natural n.*

Corollary 2. *If $A \in \mathcal{L}(H)$ is a selfadjoint operator and $P_N(t)$ is a polynomial, then $P_N(A)$ is a selfadjoint operator.*

For selfadjoint operator $D(A) = H$. There exists a wider class of linear operators that possess the properties of selfadjoint operators but are defined only on everywhere dense in H linear manifold; that is, for them $\overline{D(A)} = H$.

Linear operator A assigned on everywhere dense in H linear subspace $D(A)$, is termed *symmetric* if for any $x, y \in D(A)$ equality (8.7) holds.

Selfadjoint operator A represents a special case of the symmetric one, namely, it is a bounded symmetric operator for which $D(A) = H$. In the general case the symmetric operator can be unbounded.

Example 8.3. Let $H = L_2(0, \pi)$, $D(A) = \{u(x) : u(x) \in C^2[0, \pi], u(0) = u(\pi) = 0\}$,

$$Au = -\frac{d^2 u}{dx^2} + u.$$

Check the symmetry of operator A: let $u, v \in D(A)$, then integrate twice by parts with regard to zero values of u, v for $x = 0, \pi$ to obtain

$$\int_0^\pi \left(-\frac{d^2 u}{dx^2} + u\right) v dx = \int_0^\pi u \left(-\frac{d^2 v}{dx^2} + v\right) dx.$$

This operator, as an operator acting from $L_2(0, \pi)$ into $L_2(0, \pi)$, is unbounded. In fact, take a sequence of functions $u_n(x) = \sqrt{2/\pi} \sin nx$, $n = 1, 2, \ldots$. We have $\|u_n(x)\|_{L_2} = 1$ although it is easy to calculate that

$$\|Au_n\|^2_{L_2} = (n^2 + 1) \to \infty \quad \text{for} \quad n \to \infty.$$

Symmetric operators satisfy the statements we proved before for a selfadjoint operator under additional restrictions connected with the correspondence among the domains of definitions and the domains of values. We now formulate further some definitions and statements for symmetric operators assuming by default that if $D(A) = H$, then anything we say is also correct for selfadjoint operators.

Expressions (Ax, y) and (Ax, x) are called, respectively, *bilinear and quadrature forms*, generated by the operator A. It is possible to establish, for some pairs of symmetric operators, the relationships "greater or equal" and "greater."

Symmetric operator A is termed *nonnegative* ($A \geq 0$) if $(Ax, x) \geq 0$ for $\forall\, x \in D(A)$, $x \neq 0$. Symmetric operator A is referred to as *positive definite* ($A > 0$) if there exists a constant $c > 0$ such that $(Ax, x) \geq c(x, x)$ for $\forall\, x \in D(A)$. One may write $A \geq B$ or $A > B$, if the operator $A - B$ is a nonnegative or positive definite operator, respectively.

Lemma 2. *Let $A > 0$ be a symmetric operator; then the generalized Cauchy–Bunyakovskii inequality is valid for $\forall\, x, y \in D(A)$:*

$$|(Ax, y)| \leq (Ax, x)^{1/2}(Ay, y)^{1/2}. \tag{8.8}$$

□ Define the new inner product on $D(A)$ with the formula

$$[x, y] = (Ax, y). \tag{8.9}$$

It satisfies all axioms of Hermitian space and, consequently, the Cauchy–Bunyakovskii inequality

$$|[x, y]| \leq [x, x]^{1/2}[y, y]^{1/2}$$

holds for it, which is inequality (8.8). ■

We had introduced in Section 2 of this chapter the operator P of orthogonal projection from a space H onto some of its subspace L and investigated some of its properties; let us continue this study.

Lemma 3. *It is necessary and sufficient for linear operator P in H to be an orthoprojector, that:*

(1) *P should be a selfadjoint operator;*

(2) $P^2 = P$.

□ *Necessity.* If $x_1 = y_1 + z_1$, $x_2 = y_2 + z_2$, where $y_1, y_2 \in L$ and $z_1, z_2 \in L^{\perp}$, then

$$(Px_1, x_2) = (y_1, y_2 + z_2) = (y_1, y_2) + (y_1, z_2)$$

$$= (y_1, y_2) = (y_1 + z_1, y_2) = (x_1, Px_2).$$

The self-adjointness is proven; finally, $P^2 x_1 = P(Px_1) = Py_1 = Px_1$.

Sufficiency. Denote by L a set of all $x \in H$ for which $Px = x$; obviously, L is a subspace in H. Let x be an arbitrary element from H, then

$$x = Px + (x - Px) \tag{8.10}$$

and $Px \in L$, since $P(Px) = Px$. If $u \in L$, then $Pu = u$. Consequently,

$$(x - Px, u) = (x, u) - (Px, u) = (x, u) - (x, Pu) = 0;$$

that is, $x - Px \perp L$. Formula (8.10) thus performs orthogonal expansion of H into two subspaces [see (2.8)]. ∎

Since $(Px_1, x_1) = (y_1, y_1 + z_1) = (y_1, y_1) \geq 0$, then $P \geq 0$.

3. Norm of selfadjoint operator, its upper and lower bounds

Let us prove an important theorem for a norm of a selfadjoint operator.

Theorem 3. If an operator A is selfadjoint, then

$$\|A\| = \sup_{\|x\|=1} |(Ax, x)|. \tag{8.11}$$

□ By the Cauchy–Bunyakovskii inequality we have for $\|x\| = 1$

$$|(Ax, x)| \leq \|Ax\| \, \|x\| \leq \|A\|.$$

Consequently, if $C = \sup\limits_{\|x\|=1} |(Ax, x)|$, then $C \leq \|A\|$. Prove the inverse inequality. Notice first that substituting $x = z/\|z\|$ for $\forall\ z \in H$ into the formula for C, we obtain

$$|(Az, z)| \leq C\|z\|^2 \tag{8.12}$$

For $\forall\, x, y \in H$ we have

$$(A(x + y),\, x + y) = (Ax, x) + 2(Ax, y) + (Ay, y)$$

$$(A(x - y),\, x - y) = (Ax, x) - 2(Ax, y) + (Ay, y).$$

Subtract to obtain

$$4(Ax, y) = (A(x + y),\, x + y) - (A(x - y), x - y)$$

whence, with regard to (8.12) and the parallelogram formula, we arrive at

$$|(Ax, y)| \leq \frac{C}{4}\left(\|x + y\|^2 + \|x - y\|^2\right) \leq \frac{C}{2}\left(\|x\|^2 + \|y\|^2\right)$$

Assuming that $Ax \neq 0$ (then $x \neq 0$ too), assign in the last inequality $(\|x\|/\|Ax\|)Ax$. Then $\|y\| = \|x\|$ and we obtain

$$\frac{\|x\|}{\|Ax\|}\|Ax\|^2 \le C\|x\|^2$$

or $\|Ax\| \le C\|x\|$. This inequality is also valid for $x = 0$. Consequently, $\|A\| \le C$; that is, $\|A\| = C$. ∎

It is possible to introduce for a selfadjoint operator A the notions of *upper* M_A and *lower* m_A bounds of the operator A:

$$M_A = \sup_{\|x\|=1} (Ax, x), \qquad m_A = \inf_{\|x\|=1} (Ax, x). \tag{8.13}$$

It is easy to deduce by the definition of bounds, that for $\forall\, x \in H$ the equality

$$m_A(x,x) \le (Ax,x) \le M_A(x,x) \tag{8.14}$$

holds; that is, $m_A I \le A \le M_A I$, and that by Theorem 3

$$\|A\| = \max\left(|M_A|, |m_A|\right). \tag{8.15}$$

Therefore, if for selfadjoint operators A and B the equality $(Ax, x) = (Bx, x)$ holds for all $x \in H$, then $A = B$.

§ 9. Eigenvalues and Eigenelements of Selfadjoint and Symmetric Operators

Reality of quadrature form. Properties of eigenvalues and eigenelements of selfadjoint and symmetric operators (reality of eigenvalues and orthogonality of eigenelements). Eigenpairs of operator's power, polynomials of operator, positive definite operator. Bounds of eigenvalues of selfadjoint operator. Relay relationship. Definition of a function of selfadjoint operator.

We prove the reality of eigenvalues of symmetric operators in this section; therefore, we assume a Hilbert space H to be complex, where an inner product satisfies the condition: $(x, y) = \overline{(y, x)}$. Recall that all properties proven for a symmetric operator are also valid for a selfadjoint operator.

1. Reality of quadrature form of symmetric operator. Properties of eigenvalues and eigenelements of selfadjoint and symmetric operators, powers of operators, polynomials of operator, positive definite operators

Lemma 1. *If A is a symmetric selfadjoint operator, then the quadrature form (Ax, x), $x \in D(A)$ takes only real values.*

□ In fact, by (8.7)

$$(Ax, x) = (x, Ax) = \overline{(Ax, x)} \quad \forall x \in D(A). \blacksquare$$

Rememer that λ is called an *eigenvalue* of linear operator A if there exists an element $x \neq 0$, $x \in D(A)$ for which

$$Ax = \lambda x. \tag{9.1}$$

In this case, x is called an *eigenelement* of the operator A corresponding to the eigenvalue λ. Obviously, all elements of type αx, $\alpha \neq 0$ for the same λ are eigenelements of operator A together with x. Maximal number r $(1 \leq r \leq \infty)$ of linearly independent eigenelements, corresponding to given eigenvalue λ, is called a *multiplicity* of this eigenvalue; if $r = 1$, then λ is referred to as a *simple* eigenvalue. Multiplicity can be infinite: for instance, for the operator $A = I$ the eigenvalue $\lambda = 1$ has the multiplicity equal to a dimension of H.

Lemma 2. *If multiplicity of an eigenvalue λ of an operator A is $r > 1$ and $u_1, u_2, \ldots, u_r$ are corresponding linearly independent eigenelements, then any combination of them is also an eigenelement corresponding to this eigenvalue.*

□ Indeed, for $\forall n \geq 1$, $1 \leq n_1 < n_2 < \ldots < n_n \leq r$

$$A\left(\sum_{k=1}^{n} \alpha_k u_{n_k}\right) = \sum_{k=1}^{n} \alpha_k A u_{n_k} = \lambda \sum_{k=1}^{n} \alpha_k u_{n_k}. \blacksquare$$

Corollary 1. *It is possible to select eigenelements, corresponding to the same eigenvalue, as orthonormed.*

□ To do so, apply to the elements $u_1, u_2, \ldots, u_r, \ldots$ the Sonin–Schmidt orthogonalization process with subsequent norming of obtained orthogonal elements. ■

Theorem 1. *Eigenvalues of symmetric operator A are real and eigenelements corresponding to different eigenvalues are orthogonal.*

□ Let (λ_i, u_i) be an eigenpair of an operator A; then we have the following chain of equalities:

$$\lambda_i(u_i, u_i) = (\lambda_i u_i, u_i) = (Au_i, u_i) = (u_i, Au_i) = (u_i, \lambda_i u_i) = \bar{\lambda}_i(u_i, u_i);$$

that is, $\lambda_i = \bar{\lambda}_i$, and therefore, λ_i is real. Let (λ_j, u_j) be another eigenpair of the operator A and let $\lambda_i \neq \lambda_j$ be real; then we obtain the chain of equalities:

$$\lambda_i(u_i, u_j) = (\lambda_i u_i, u_j) = (Au_i, u_j) = (u_i, Au_j)$$
$$= (u_i, \lambda_j u_j) = \lambda_j(u_i, u_j). \quad (9.2)$$

Hence, since $\lambda_i \neq \lambda_j$, it follows that $(u_i, u_j) = 0$. ■

Assume that the set of all linearly independent eigenelements of a symmetric operator A is not more than countable. Let us renumber all its eigenvalues: $\lambda_1, \lambda_2, \ldots$, repeating λ_k as many times as equal to its multiplicity. Denote corresponding eigenelements by $u_1, u_2, \ldots, u_k, \ldots$ so that each eigenvalue would have one corresponding eigenelement u_k:

$$Au_k = \lambda_k u_k, \quad k = 1, 2, \ldots . \quad (9.3)$$

By Lemma 2, Corollary 1, and Theorem 1 the following corollary is valid.

Corollary 2. *If a system of all linearly independent eigenelements of symmetric operator A is not more than countable, then it may be chosen as an orthonormed one:*

$$(Au_k, u_i) = \lambda_k(u_k, u_i) = \lambda_k \delta_{ki}. \quad (9.4)$$

Lemma 3. *If (λ_k, u_k) is an eigenpair of an operator A, then for any natural n, (λ_k^n, u_k) is an eigenpair of the operator A^n.*

□ Indeed,

$$A^n u_k = A^{n-1}(Au_k) = \lambda_k A^{n-1} u_k = \ldots = \lambda_k^n u_k$$

and $u_k \in D(A^n)$. ■

Corollary 3. *If $P_N(t)$ is a polynomial of Nth power, then $(P_N(\lambda_k), u_k)$ is an eigenpair of the operator $P_N(A)$.*

Lemma 4. *Eigenvalues of positive definite symmetric operator A are positive.*

□ We have

$$c(u_k, u_k) \leq (Au_k, u_k) = \lambda_k(u_k, u_k);$$

that is, $\lambda_k \geq c > 0$. ■

2. Bounds of eigenvalues of selfadjoint operator. Relay relationship. Definition of a function of selfadjoint operator

Let A be a selfadjoint operator, and let $A \in \mathcal{L}(H)$, (λ_k, u_k) be an eigenpair of operator A. Set $x = u_k$ in (8.14) to obtain the bounds for eigenvalues:

$$m_A \leq \lambda_k \leq M_A. \tag{9.5}$$

The following theorem answers the question of when the extreme number in inequality (9.5) belongs to eigenvalues.

Theorem 2. *It is necessary and sufficient for (M_A, φ) to be an eigenpair of selfadjoint operator A, that $(A\varphi, \varphi) = M_A(\varphi, \varphi)$. A similar statement is valid for (m_A, φ).*

□ *Necessity.* If M_A and φ are such that $A\varphi = M_A\varphi$, then

$$(A\varphi, \varphi) = M_A(\varphi, \varphi).$$

Sufficiency. Let $(A\varphi, \varphi) = M_A$ for some $\varphi \in H$ with $\|\varphi\| = 1$; that is, $(A\varphi - M_A\varphi, \varphi) = 0$. But, since for $\forall\, v \in H$,

$$M_A(v, v) - (Av, v) = (M_A v - Av, v) \geq 0$$

then we have for any real t and $\forall\, \eta \in H$

$$(M_A(\varphi + t\eta) - A(\varphi + t\eta), \varphi + t\eta) \geq 0.$$

Open this inequality to get

$$(M_A\varphi - A\varphi, \varphi) + t(M_A\eta - A\eta, \varphi) + t(M_A\varphi - A\varphi, \eta)$$
$$+ t^2(M_A\eta - A\eta, \eta) \geq 0.$$

The first member in the last inequality is equal to zero, the second and the third are equal, and the last one is nonnegative; therefore,

$$(M_A\varphi - A\varphi, \eta) = 0 \quad \text{for} \quad \forall\, \eta \in H;$$

that is, φ is an eigenelement and $A\varphi = M_A\varphi$. ■

Let $x \neq 0$, and denote

$$\Phi_A(x) = \frac{(Ax, x)}{(x, x)}; \tag{9.6}$$

this relationship is called the *Relay relationship.* By inequality (8.14) for $\forall\, x \neq 0$, $x \in H$,

$$m_A \leq \Phi_A(x) \leq M_A. \tag{9.7}$$

If the conditions of Theorem 2 are satisfied, we obtain the formulas for maximal and minimal eigenvalues:

$$\max_i \lambda_i = \sup_{\substack{x \in H \\ x \neq 0}} \frac{(Ax, x)}{(x, x)}, \qquad \min_i \lambda_i = \inf_{\substack{x \in H \\ x \neq 0}} \frac{(Ax, x)}{(x, x)}. \tag{9.8}$$

Now let us take a strong assumption on a spectrum of selfadjoint operator A.

Assumption. Let a system of eigenelements $\{u_k\}$ of selfadjoint operator A be not more than countable and make up an orthonormed basis in the space H. We then say that operator A has a *purely pointwise spectrum* and is an operator of *scalar type.*

We can now prove formula (4.5) for the spectral radius of selfadjoint operator A:

$$\mu(A) = \sup_{\lambda \in \mathrm{Sp}(A)} |\lambda|. \tag{9.9}$$

□ In fact, by Theorem 2 from Section 4 of this chapter $\mu(A) = \|A\|$ and by (8.15)

$$\|A\| = \sup_{\substack{x \in H \\ x \neq 0}} |\Phi_A(x)|.$$

We obtain now, with regard to relationships (9.8), formula (9.9). ■

If the system $\{u_k\}$ is complete and orthonormed in H, then it is possible to expand $\forall\, x \in H$ into the Fourier series:

$$x = \sum_{k=1}^{\infty} x_k u_k, \qquad x_k = (x, u_k). \tag{9.10}$$

Under these conditions, the action of operator A is expressed with the formula

$$Ax = \sum_{k=1}^{\infty} \lambda_k x_k u_k. \tag{9.11}$$

Consequently, if a function $f(x)$ is defined on some set Ω that includes all eigenvalues of the selfadjoint operator $A(\lambda_k \in \Omega)$, then it is possible to express the operator $f(A)$ (its action to the element x), if the corresponding series converges, with the formula

$$f(A)x = \sum_{k=1}^{\infty} f(\lambda_k) x_k u_k. \tag{9.12}$$

For example,

$$\exp(A)x = \sum_{k=1}^{\infty} \exp(\lambda_k) x_k u_k, \tag{9.13}$$

or, for $\lambda_k \geq 0$

$$A^{1/2}x = \sum_{k=1}^{\infty} \sqrt{\lambda_k} x_k u_k \tag{9.14}$$

and for $\lambda_k \neq 0$

$$A^{-1}x = \sum_{k=1}^{\infty} \lambda_k^{-1} x_k u_k. \tag{9.15}$$

Having known a basis $\{u_k\}$ of eigenelements of an operator A, we can find a solution to the equation

$$Ax = f, \tag{9.16}$$

$f \in H$, applying the following algorithm. Let λ_k be the eigenvalue corresponding to an element u_k and let all $\lambda_k \neq 0$. Let us expand f into the series

$$f = \sum_{k=1}^{\infty} f_k u_k \tag{9.17}$$

and search for a solution of Equation (9.16) in the form (9.10). Substitute (9.10) and (9.17) into (9.16) to obtain

$$A\sum_{k=1}^{\infty} x_k u_k = \sum_{k=1}^{\infty} x_k A u_k = \sum_{k=1}^{\infty} \lambda_k x_k u_k = \sum_{k=1}^{\infty} f_k u_k.$$

The elements u_k are linear independent; therefore, equate the expressions at u_k to obtain

$$\lambda_k x_k = f_k, \quad k = 1, 2, \ldots .$$

Hence, $x_k = f_k/\lambda_k$ and, consequently, a solution of Equation (9.16) has the form

$$x = \sum_{k=1}^{\infty} \lambda_k^{-1} f_k u_k = A^{-1} f. \tag{9.18}$$

§ 10. Quadrature Functionals with Positive Definite Symmetric or Symmetrizable Operator and Generalized Solutions of Operator Equations

Quadrature functionals with positive definite symmetric operator. Energetic space, reduction of a functional to canonical form, functionals of errors, energy, least squares, and generalized least squares. Theorem on minimization of a functional of energy and solution of equation with selfadjoint operator. Notion on generalized solution. Properties of minimizing sequences, symmetrizable operators.

1. Quadrature functionals with positive definite symmetric operator. Energetic space. The reduction of a functional to canonical form. Functionals of errors, energy, least and generalized least squares

Let H be real Hilbert space with an inner product $(\cdot,\cdot)$ and a norm $\|\cdot\|$ and let M be a symmetric positive definite operator, $D(M) \subset H$, $\overline{D(M)} = H$, $l(x)$ be a linear bounded functional, $D(l) = H$. Take a quadrature functional of the form

$$G_M(x) = (Mx, x) - 2l(x) + C_0 \tag{10.1}$$

defined on $x \in D(M)$ where C_0 is a constant. We now try to find

$$\inf G_M(x) \tag{10.2}$$

and an element u (if it exists) that realizes (10.2):

$$u = \arg\inf G_M(x).$$

The operator M may be defined not on all $x \in H$; therefore, let us perform the following construction of new energetic space according to the technique developed before (see Section 7 of Chapter 1).

Define on elements $x, y \in D(M)$ a new inner product $[\cdot,\cdot]$ and a norm $[\cdot]_M$:

$$[x, y] = (Mx, y), \qquad [x]_M^2 = (Mx, x). \tag{10.3}$$

Let $D(M) \neq H$. Completing $D(M)$ with respect to the norm $[\cdot]_M$ we obtain a new space H_M with inner product (10.3). Due to positive definiteness of the operator M : $(Mx, x) \geq c(x, x)$ for $\forall x \in D(M)$ we obtain that $[x]_M^2 \geq c\|x\|^2$. The linear functional $l(x)$ being bounded on $D(M)$ is bounded on this set in the space H_M too, since $|l(x)| \leq \|l\| \cdot \|x\| \leq c^{-1}\|l\|[x]_M$. Extend it with respect to continuity from everywhere dense in H_M set $D(M)$ to all space H_M. Then, by the Riesz theorem, there exists a unique element $u \in H_M$ such that $l(x) = [u, x]_M$, and the quadrature functional $G_M(x)$ takes the form

$$G_M(x) = [x, x]_M - 2[u, x]_M + C_0 = [x - u]_M^2 - [u]_M^2 + C_0. \tag{10.4}$$

For $D(M) = H$ the space H_M will consist of the elements of the space H.

Formula (10.4) shows that $G_M(x)$ attains the least value equal to $-[u]_M^2 + C_0$ for unique value $x = u$. The quadrature functional

$$\Omega_M(x) = [x - u, x - u]_M \tag{10.5}$$

is called an *error functional*; its minimum equals zero and is reached for $x = u$. Let us present examples of quadrature functionals used while solving the equation

$$Au = f. \tag{10.6}$$

For $M = A$ with symmetric positive definite operator A:

$$G_A(x) = (Ax, x) - 2(x, f). \tag{10.7}$$

For $M = A^*A$:

$$G_{A^*A}(x) = (Ax - f, Ax - f). \tag{10.8}$$

For $M = A^*BA$ where B is a symmetric positive definite operator:

$$G_{A^*BA}(x) = (B(Ax - f), Ax - f). \tag{10.9}$$

The functional (10.7) is called an *energy functional* and (10.8) and (10.9) are *functionals of least* (*generalized least*) *squares*; we have selected, due to the choice of the constant C_0, the form of the last functionals so that they coincide with corresponding error functionals whose minimum is attained for $x = u$; the operator A in (10.8) and (10.9) is not necessarily symmetric.

2. Theorem on the minimization of energy functional and Solution of equation with selfadjoint operator. Notion on generalized solution

As we said already, the minimum of the functionals of type (10.8) and (10.9) can be guessed: it is equal to zero and attained on an element u connected in some sense with a solution of Equation (10.6). Let us find out what is the equation whose solution is connected with the element realizing a minimum of energy functional (10.7). Let A be the selfadjoint positive definite operator; the following important theorem is valid for it.

Theorem 1. *It is necessary and sufficient for some element from* $u \in H$ *to provide minimal value to the energy functional* (10.7) *with selfadjoint positive definite operator* A, *that this element satisfies Equation* (10.6).

□ *Sufficiency.* Let $Au = f$; take $x = u + \eta$ where η is an arbitrary element from H; then it easy to check that

$$G_A(x) = G_A(u) + (A\eta, \eta) + 2(Au - f, \eta)$$

$$= G_A(u) + (A\eta, \eta) \geq G_A(u) \tag{10.10}$$

since $(A\eta, \eta) \geq 0$.

Necessity. Let u realize a minimum and let $\eta \in H$, $\eta \neq 0$. Take an element $u + t\eta$ for any t and calculate

$$G_A(u + t\eta) = G_A(u) + 2t(Au - f, \eta) + t^2(A\eta, \eta)$$

and

$$\left.\frac{\partial G_A}{\partial t}\right|_{t=0} = 2(Au - f, \eta) = -(r, \eta), \tag{10.11}$$

where $r = f - Au$ is a discrepancy.

The following equation holds in a point of minimum

$$(Au - f, \eta) = 0 \quad \text{for} \quad \forall \eta \in H. \tag{10.12}$$

Set $\eta = Au - f$ to obtain that the element u must be a solution to Equation (10.6). ■

What do we do, if in (10.7) the operator A is not selfadjoint and is only symmetric positive definite? In this case the functional has a minimum too, although on the element $u \in H_A$ and it may happen that $u \notin D(A)$. In this case the element u is called a generalized solution of Equation (10.6). More often, which is more convenient for many problems, equality (10.12) in a somewhat different formis is used for the definition of the generalized solution. We call the *generalized solution* to Equation (10.6) an element $u \in H_A$ such that for any $\eta \in H_A$ the following equality holds

$$[u, \eta]_A = (f, \eta) \quad \text{for} \quad \forall\, \eta \in H_A. \tag{10.13}$$

We see, using equality (10.10), that the element u, satisfying Equation (10.13), provides a minimum to the energy functional. Equations of the type (10.13) are termed, because of their connection with the functional minimization problem, *variational.* So, it is important to understand that the functional minimization problem of type (10.1) and (10.4) on specially constructed energetic space H_M and the problem of the search for the generalized solution of Equation (10.6) with a symmetric positive definite operator are equivalent problems.

3. Properties of minimizing sequences

A sequence $\{u_n\} \in H_M$ is referred to as *minimizing* a quadrature functional $G_M(x)$ if

$$G_M(u_n) \to \inf G_M(x) = G_M(u), \qquad u \in H_M.$$

Theorem 2. *If a sequence $\{u_n\}$ is minimizing, then $u_n \to u$ in a norm of the space H_M.*

□ We have

$$G_M(u_n) = [u_n - u]_M^2 - [u]_M^2 + C$$

and

$$G_M(u) = -[u]_M^2 + C,$$

consequently,

$$[u_n - u]_M^2 = G_M(u_n) - G_M(u) \to 0$$

for $n \to \infty$. ■

Algorithms for the construction of minimizing sequences are presented in Section 11 of this chapter (the generalized Ritz method) and in Section 7 of Chapter 3 (the descent methods).

4. Symmetrizable operators, quadrature functionals

It is possible to transfer many methods of investigation and justification of algorithms for the solution of problems with a selfadjoint operator to problems with symmetrizable operators, performing substitutions of variables and operators.

Let H be a Hilbert space with inner product $(\cdot,\cdot)$, $\mathbf{X}$, $\mathbf{Y}$ be Banach spaces, operator $L \in L(\mathbf{X},\mathbf{Y})$, and let H_M be a Hilbert space with inner product $[\cdot,\cdot]_M = (M\cdot,\cdot)$. Let us refer to the operator L as (C,D)-*symmetrizable* in H_M by a pair of operators C, D if operator $C \in \mathcal{L}(\mathbf{Y}, H_M)$ and operator $D \in \mathcal{L}$, that has an inverse one, will be found such that the operator $A = CLD^{-1} \in \mathcal{L}(H_M, H_M)$ is selfadjoint (and, consequently, bounded) positive definite in the H_M operator.

Take an equation for the definition of an element $v \in \mathbf{X}$

$$Lv = g, \tag{10.14}$$

where $g \in \mathbf{Y}$. If L is (C,D)-symmetrizable in the H_M operator, then acting from the left to the left and to the right sides of Equation (10.14) by the operator C and substituting a variable with the formula $u = Dv$, we obtain for $x \in H_M$ a new equation of type (10.6) with

$$A = CLD^{-1}, \qquad f = Cg. \tag{10.15}$$

Therefore, if some algorithm is known and justified (e.g., a variational or iterative one) for the solution of Equation (10.6), then, by inverse substitution of variables in its formulas with the operator D^{-1}, we obtain the corresponding algorithm for immediate solution of Equation (10.14).

So, for example, it follows by formulas (10.7) and (10.9) for quadrature functionals $G_A(x)$, $G_{A^*BA}(x)$ for (C, D)-symmetrizable in the H operator L, that the solution of Equation (10.14) realizes a minimum, respectively, for the functionals

$$\tilde{G}_1(x) = (CLx, MDx) - 2(MDx, Cg) \tag{10.16}$$

$$\tilde{G}_2(x) = (BC(Lx - g), MC(Lx - g)). \tag{10.17}$$

§ 11. Variational Methods for the Minimization of Quadrature Functionals

Generalized Ritz method, equations of the method, properties of approximate solution, convergence. Concrete realizations of the method: Ritz method, method of least squares. Notion on projection methods and Galerkin method. Remarks on the choice of coordinate elements.

There exist various methods to find a minimum for the quadrature functional (10.1). We present one of them here; it may be regarded as the generalized Ritz method for the search for the approximate (generalized) solution of Equation (10.6). We preserve in this section the notation and assumptions of the receding section.

1. The generalized Ritz method, equations of the method, properties of approximate solution. Convergence

It is efficient to write the systems of equations that form a minimizing sequence for the functional (10.1) in these forms: one, convenient for the study, and another, adapted for calculations. To this end, we also write the quadrature functional in two forms

$$G_M(x) = (Mx, x) - 2l(x) + C \tag{11.1}$$

and

$$G_M(x) = [x - u]_M^2 - [u_0]_M^2 + C. \tag{11.2}$$

Let

$$\Omega_M(x) = [x - u]_M^2. \tag{11.3}$$

Let $\{\varphi_k\}_1^\infty$ be complete in the H_M linearly independent system of elements, $n \geq 1$, and let L_n be the linear hull formed by the first n elements $\varphi_1, \varphi_2, \ldots, \varphi_n$; that is, $u_n \in L_n$ if

$$u_n = \sum_{i=1}^{n} a_i \varphi_i. \tag{11.4}$$

Substitute the problem $\min_{x \in L_n} G_M(x)$ for the problem on the search for $\min_{x \in H_M} G_M(x)$ with the hope that for large n, solutions of these problems will be close enough. Let us show that the new problem is simpler and can be reduced to the solution of a system of linear algebraic equations of nth order. In fact, substitute u_n in the form (11.4) into G_M to obtain

$$G_M(u_n) = G(a_1, a_2, \ldots, a_n),$$

where $G(a_1, \ldots, a_n)$ is the fully definite number function of parameters

$a_1, a_2, \ldots, a_n$. In its point of minimum

$$\frac{\partial}{\partial a_k} G(a_1, a_2, \ldots, a_n) = 0, \qquad k = \overline{1, n} \tag{11.5}$$

must be valid.

The system of Equations (11.5) for the functional in the form (11.1) looks like

$$(M\varphi_k, u_n) + (Mu_n, \varphi_k) - 2l(\varphi_k) = 0$$

or

$$(Mu_n, \varphi_k) = l(\varphi_k), \qquad k = \overline{1, n}.$$

Substituting here u_n from (11.4) we obtain for the determination of a_i the system of algebraic equations

$$\sum_{i=1}^{n} (M\varphi_i, \varphi_k) a_i = l(\varphi_k), \qquad k = \overline{1, n} \tag{11.6}$$

or

$$\sum_{i=1}^{n} [\varphi_i, \varphi_k]_M a_i = l(\varphi_k), \qquad k = \overline{1, n}. \tag{11.7}$$

The same system of equations for the functional in the form (11.2) looks like

$$[u_n - u, \varphi_k]_M = 0, \qquad k = \overline{1, n}. \tag{11.8}$$

The system of Equations (11.6) is called the *system of equations of the generalized Ritz method* for approximate minimization of quadrature functional (11.1) or the search for approximate solution of Equation (10.6).

We can make a number of statements on systems (11.6)–(11.8).

Lemma 1. *The system of Equations* (11.6) *always has a solution; more than that, it is unique.*

□ It follows from the fact that equivalent system (11.7) is a system with the Gram matrix composed of inner products of linearly independent in H_M elements $\{\varphi_i\}_1^n$. ■

Theorem 3. *Element u_n of type* (11.4), *where $\{a_i\}_1^n$ is a solution to system* (11.6), *is an element of the best approximation for u in the space H_M among all elements of the subspace L_n.*

□ It is sufficient (see Corollary 2, Section 8 of Chapter 1) to prove that $[u_n - u, \eta]_M = 0$ for $\forall \eta \in L_n$. Let $\eta \in L_n$; then there exists a linear combination of elements $\{\varphi_i\}_1^n$ such that $\eta = \sum_{k=1}^{n} b_k\varphi_k$. Having multiplied the kth equality (11.8) by b_k and summing, we obtain $[u_n - u, \eta]_M = 0$. ■

Corollary 1. *For any $v \in L_n$ we have*

$$[u - u_n]_M \leq [u - v]_M. \tag{11.9}$$

Corollary 2. *An error of the generalized Ritz method is orthogonal in H_M to the subspace L_n.*

Let us prove the convergence of the generalized Ritz method.

Theorem 4. *A sequence of approximations u_n obtained with the generalized Ritz method converges in the space H_M to the generalized solution of Equation* (10.6), *that provides a minimum in H_M to quadrature functional* (11.1).

□ Choose any $\varepsilon > 0$. By virtue of completeness of $\{\varphi_k\}_1^\infty$, there will be found n_0 and coefficients $b_1, b_2, \ldots, b_{n_0}$ in the space H_M such that $[u - v]_M < \varepsilon$ where $v = \sum_{k=1}^{n_0} b_k\varphi_k$; then by inequality (11.9), $[u - u_n] < \varepsilon$ for $n \geq n_0$. ■

2. Concrete realizations of the generalized Ritz method: the Ritz methods, the method of least squares. Notion on projection methods and the Galerkin method. Remarks on the choice of coordinate elements

We have already obtained equations of the generalized Ritz method, minimizing on finite-dimensional space L_n a quadrature functional. We can obtain equations of the classic Ritz method in the framework of this approach if we take the energy functional (10.7) for $M = A$ as the quadrature functional; then $l(x) = (x, f)$, $C = 0$, and a system of equations of the Ritz method takes the form:

$$\sum_{i=1}^{n}(A\varphi_i, \varphi_k)a_i = (f, \varphi_k), \qquad k = \overline{1, n} \tag{11.10}$$

or

$$(Au_n - f, \varphi_k) = 0, \qquad k = \overline{1, n}. \tag{11.11}$$

For $M = A^*A$ where $A \in \mathcal{L}(H)$ we obtain equations of the method of least squares

$$\sum_{i=1}^{n}(A\varphi_i, A\varphi_k)a_i = (A\varphi_k, f), \qquad k = \overline{1, n} \tag{11.12}$$

or

$$(Au_n - f, A\varphi_k) = 0, \qquad k = \overline{1, n}. \tag{11.13}$$

For $M = A^*BA$ where $A \in \mathcal{L}(H)$, and B is a selfadjoint operator we arrive at the method of generalized least squares

$$\sum_{i=1}^{n}(BA\varphi_i, A\varphi_k)a_i = (BA\varphi_k, f), \qquad k = \overline{1, n} \tag{11.14}$$

or

$$((Au_n - f), BA\varphi_k) = 0, \qquad k = \overline{1, n}. \tag{11.15}$$

Examples (11.10)–(11.15) demonstrate that the various Ritz methods for the finding of approximate solution to the equation

$$Au = f \tag{11.16}$$

differ from each other within admissible assumptions on problem (11.16): first, the choice of the operator M and second, the choice of a system of coordinate elements which are used to expand the approximate solution with formula (11.4).

Formulas (11.11), (11.13), and (11.15) show that the discrepancy $r_n = f - Au_n$ is always orthogonal to some finite-dimensional subspace: in (11.11) it is L_n, in (11.13) and (11.15) – subspaces, that are linear hulls of elements $\{A\varphi_k\}_1^n$ and $\{BA\varphi_k\}_1^n$, respectively. It suggests applying the following formal algorithm for the construction of an approximate solution to Equation (11.16) where the operator $A \in \mathcal{L}(H)$ is not necessarily a symmetric one.

Take, together with the subspace L_n, a subspace $\tilde{L}_n$ constructed similarly on another complete linearly independent system of elements $\{\psi_k\}_1^\infty$. Substitute into Equation (11.16) the element u_n, formed by formula (11.4) and require that the discrepancy r_n be orthogonal to the subspace $\tilde{L}_n$. This condition leads to the system of equations

$$(Au_n - f, \psi_k) = 0, \qquad k = \overline{1, n}$$

or (11.17)

$$\sum_{i=1}^{n} (A\varphi_i, \psi_k) a_i = (f, \psi_k), \qquad k = \overline{1, n}.$$

The method we have just described is called the *Galerkin method* and the system of Equations (11.17), *equations of the Galerkin method.* This method, suggested in 1915, became a precursor to the theory of generalized solutions [see formulas (10.12), (10.13), (15.4), and (15.12)]. As we have seen, the method does not imply the existence of some quadrature functional. Of course, while justifying the method one should impose some restrictions on the operator of problem (11.16).

It is possible to obtain a generalization of the method on the basis of the algorithm for the finding of systems of equations of the Galerkin method where the spaces L_n and $\tilde{L}_n$ are assigned explicitly with elements $\{\varphi_k\}_1^n$ and $\{\psi_k\}_1^n$. Let L_n and $\tilde{L}_n$ be two n-dimensional subspaces in H and let $\tilde{P}$ be an orthoprojector from H in $\tilde{L}_n$; then Equation (11.16) can be substituted with an approximate one, such as

$$\tilde{P}Au_n = \tilde{P}f, \qquad u_n \in L_n. \tag{11.18}$$

Equation (11.18) (as a matter of fact, it is a system of n linear algebraic equations) is an equation of the so-called *projection method.* Notice that equations of the Galerkin method are equations of type (11.18).

We now make a number of concluding remarks on the choice of coordinate elements.

Remark 1. In some variants of the preceding methods for the construction of approximations for Equations (11.16), the spaces L_n and $\tilde{L}_n$ form linear hulls of elements, respectively, in the forms $\{\varphi_1, B_1\varphi_1, \ldots, B_1^{n-1}\varphi_1\}$ and $\{\psi_1, B_2\psi_1, \ldots, B_2^{n-1}\psi_1\}$ under special assumptions about the operators $B_1, B_2 \in \mathcal{L}(H)$. These variants are called the *momentum methods.*

Remark 2. While numerically realizing the preceding methods, one must solve systems of equations of high order. Therefore, in order to ensure the stability of the algorithm for this system when its order n grows higher, one should take care that the condition measure for a matrix of the system of equations in question not be spoiled too much with growing n.

Remark 3. The so called *variational-difference, projection-difference, finite element*, and like methods are widely used in problems of mathematical physics. The specific character of these methods is in the fact that coordinate elements there are finite functions constructed in a special way. As a result, a matrix of the system of equations to be solved is sparse which permits the successful applcation of iteration methods.

§ 12. Variational Equations. The Vishik–Lax–Milgram Theorem

Bilinear forms, continuity, V-ellipticity. Variational equations. Vishik–Lax–Milgram theorem on existence, uniqueness, and continuous dependence of solutions of variational equations. Symmetric case, connection with the problem on a minimum for a quadrature functional.

The material of this section has proven to be an extremely important tool for the investigation of aspects of existence and uniqueness of generalized solutions of boundary value problems for linear partial differential equations of elliptic type.

For convenience, we denote here real Hilbert space by the letter V, inner product by $[\cdot,\cdot]$, a norm by $|\cdot|$, and elements of V by the letters $u, v \ldots$.

1. Bilinear forms, continuity, V-ellipticity. Variational equations

We say that *bilinear form* $a(u,v)$ is assigned on elements of Hilbert space V if any pair of elements $u, v \in V$ has corresponding real number $a(u,v)$ and this correspondence possesses the following properties: for any $u, u_1, u_2, v, v_1, v_2 \in V$ and real scalars λ_1, λ_2, μ_1, μ_2

$$\begin{aligned} a(\lambda_1 u_1 + \lambda_2 u_2, v) &= \lambda_1 a(u_1, v) + \lambda_2 a(u_2, v) \\ a(u, \mu_1 v_1 + \mu_2 v_2) &= \mu_1 a(u, v_1) + \mu_2 a(u, v_2). \end{aligned} \tag{12.1}$$

The bilinear form is termed *symmetric* if for $\forall u\, v \in V$

$$a(u,v) = a(v,u). \tag{12.2}$$

A *quadrature functional* $a(u,u)$ assigned on V is called a *quadrature form* of bilinear form $a(u,v)$. A bilinear form is termed *bounded* if there exists a positive constant $\delta > 0$ such that for $\forall\, u \in V,\ v \in V$

$$|a(u,v)| \le \delta |u| \cdot |v|. \tag{12.3}$$

A bilinear form is termed *V-elliptic* if there exists a positive constant γ such that for $\forall\, u \in V$

$$a(u,u) \ge \gamma |u|^2. \tag{12.4}$$

The inner product $[u,v]$ is an example of a symmetric bounded V-elliptic bilinear form; its quadrature form is a square of a norm.

Symmetric bounded V-elliptic bilinear form $a(u,v)$ may be regarded as a new inner product in V. Then a new norm is defined with the equality $|u|' = a(u,u)^{1/2}$ and by inequalities (12.3) and (12.4), the generelized Cauchy–Bunyakovskii inequality follows

$$|a(u,v)| \le \delta\gamma^{-1}(a(u,u))^{1/2}(a(v,v))^{1/2}. \tag{12.5}$$

Let $f(v)$ be some linear bounded functional assigned on elements $v \in V$. Take an equation called a *variational equation* to find an element $u \in V$ satisfying the inequality

$$a(u,v) = f(v) \quad \text{for} \quad \forall\, v \in V. \tag{12.6}$$

We can see that the element v in Equation (12.6) is a varying element whereas the element u is a fixed desired element. Equation (10.13) is an example of such an equation; it defines the generalized solution of Equation (10.6) with symmetric operator A. The name "variational" is related to the fact that in the case of the symmetric V-elliptic bilinear form, Equation (12.6) is equivalent to the problem on minimization of some quadrature functional. Being applied to differential equations, (12.6) has a form of some integral identity that defines generalized solution (see Section 15 of this chapter).

Theorem 1 (by Vishik–Lax–Milgram). Let V be a Hilbert space with inner product $[\cdot,\cdot]$, and let $a(u,v)$ be the bilinear form on V for which inequalities (12.3) and (12.4) *hold. Then each linear continuous on V functional $f(v)$ is represented in the form $f(v) = a(u,v)$ with some $u \in V$. Moreover, the functional f is uniquely representable and there exists a uniquely determined bounded linear continuously reversible operator A such that $a(u,v) = [Au, v]$, $\|A\| \le \delta$, $\|A^{-1}\| \le \gamma^{-1}$.*

□ If u is fixed, then the functional $a(u,v)$ is bounded on V. Therefore, by the Riesz theorem, there exists a unique element $w = Au \in V$ such that $a(u,v) = [Au, v]$. It is obvious that A is the linear operator on V and by (12.4) $R(A) = V$. Use (12.3) to obtain

$$|Au|^2 = [Au, Au] = a(u, Au) \le \delta |Au|\,|u|;$$

that is, $|Au| \le \delta|u|$ which means that operator A is bounded and $\|A\| \le \delta$. On the other hand, by (12.4),

$$|u|^2 \le \gamma^{-1} a(u,u) = \gamma^{-1}[Au, u] \le \gamma^{-1}|Au|\,|u|;$$

that is, $|u| \le \gamma^{-1}|Au|$ and by Theorem 1 from Section 3 of this chapter we obtain that operator A^{-1} exists and $\|A^{-1}\| \le \gamma^{-1}$. Consequently, the mapping $A : V \to V$ is one-to-one.

If f is a linear continuous functional then, as we mentioned, there exists a unique element $w \in V$ for which $f(v) = [w, v]$. If $Au = w$, then

$$f(v) = [Au, v] = a(u, v) \tag{12.7}$$

with $|f| = |w| \le \delta|u|$ and $|u| \le \gamma^{-1}|f|$. ■

Notice, that by the preceding proof of the theorem it is possible for each variational equation (12.6) to find a certain corresponding equation

$$Au = w, \tag{12.8}$$

where the operator A is determined by the type of bilinear form $a(u,v)$ and w by the form of linear functional $f(v)$.

2. Symmetric case. Connection with the problem of a minimum for a quadrature functional

Let $a(u,v) = a(v,u)$ for $\forall u\, v \in V$. Take a quadrature functional in the form

$$G(v) = a(v,v) - 2f(v), \tag{12.9}$$

where $f(v)$ is some bounded linear functional for $\forall\, v \in V$ and bilinear form $a(u,v)$ is bounded and V-elliptic. Let M be a closed convex set from $V(M \subset V)$. Let us pose the following *abstract minimization problem*: find an element $\tilde{u}$ such that

$$\tilde{u} = \arg \inf_{v \in M} G(v). \tag{12.10}$$

Show that the posed problem is reduced to the problem of type I we investigated in Chapter 1 on the best approximation of a given element in Hilbert space by the elements of a convex set. In fact, by (12.7) and (12.1),

$$G(v) = a(v,v) - 2a(u,v) = a(v-u, v-u) - a(u,u).$$

Consequently, the solution of the abstract minimization problem is equivalent to the minimization of a distance between an element u and a set M as related to the norm $\|\cdot\| = (a(\cdot,\cdot))^{1/2}$ formed by the new inner product $[\cdot,\cdot]' = a(\cdot,\cdot)$. Therefore, according to the results of Section 8 of Chapter 1, a solution to problem (12.10) is just a projection of the element u to the set M as related to the inner product $a(\cdot,\cdot)$. If $u \in M$, then an element v realizing (12.10) is equal to u and by (12.7) u is a solution to the variational equation (12.6); that is, $\tilde{u} = u$. If $u \notin M$, then such an element u does exist and is unique (see Theorem 1, Section 8 of Chapter 1). Theorem 2 of the same section provides the necessary and sufficient conditions determining this element. This theorem in an equivalent statement as applied to the problem (12.10) has the following form.

Theorem 2. *An element $\tilde{u}$ is a solution to the abstract minimization problem* (12.10) *for $u \notin M$ iff it satisfies the relationships*

$$\tilde{u} \in M \quad \text{and for} \quad \forall\, v \in M \quad a(\tilde{u}, v - \tilde{u}) \geq f(v - \tilde{u}) \tag{12.11}$$

in the general case and

$$\tilde{u} \in M, \qquad \forall\, v \in M \qquad a(\tilde{u}, v) = f(v) \tag{12.12}$$

if M is closed space.

□ By Theorem 2 (Section 8 of Chapter 1) the projection $\tilde{u}$ is completely characterized by the relationship

$$\tilde{u} \in M \quad \text{and for} \quad \forall\, v \in M \quad a(u - \tilde{u}, v - \tilde{u}) \leq 0.$$

Write these inequalities as

$$\forall\, v \in M \quad a(\tilde{u}, v - \tilde{u}) \geq a(u, v - \tilde{u}) = f(v - \tilde{u})$$

and the relationships (12.11) are proven. If M is a subspace, then we obtain by Corollary 2 on a projection (Section 8 of Chapter 1) that

$$a(u - \tilde{u}, v - \tilde{u}) = 0;$$

that is,

$$a(\tilde{u}, v - \tilde{u}) = a(u, v - \tilde{u}) = f(v - \tilde{u})$$

for $\forall\, u \in V$. After the substitution $\eta = v - \tilde{u}$ we have with accuracy up to notations the equality (12.12). ■

A bit later using the theorems of this section, we investigate generalized solutions of the first boundary value problem for an elliptic equation of the second order (see Section 15 of this chapter).

§ 13. Compact (Completely Continuous) Operators in Hilbert Space

Some properties of compact sets in Hilbert space. Compact operators and their properties. Theorem on representation of compact operator. Equation of the second kind with compact operator. Formulation of Fredholm theorems. Eigenvalues and eigenelements of compact operator.

1. Some properties of compact sets in Hilbert space

The definition of compact sets in a metric space was presented in Section 2 of Chapter 1. We set forth here the necessary knowledge on compact sets lying in Hilbert space H. We have proven before that any compact set is bounded, although the inverse statement is, generally speaking, incorrect. Anyhow, for finite-dimensional spaces the following lemma is valid.

Lemma 1. *If a set M from finite-dimensional (n-dimensional) Hilbert space H is bounded, it is necessary and sufficient for it to be compact.*

□ Let $e_1, e_2, \dots, e_n$ be an orthonormed basis in H. By virtue of boundedness of M, $\|f\| < C$ for $\forall\, f \in M$. Therefore, the Fourier coefficients $f_i = (f, e_i)$, $i = \overline{1,n}$ in the expansion $f = \sum_{i=1}^{n} f_i e_i$ of arbitrary element $f \in M$ satisfy the inequalities $|f_i| \le \|f\| < C$. Consequently, for any sequence of elements f^k, $k = 1, 2, \dots$, from M a sequence of n-dimensional vectors $(f_1^k, \dots, f_n^k)$, where $f_i^k = (f^k, e_i)$, is bounded. By the Boltsano–Weierstrass theorem it is possible to select from it a fundamental sequence $(f_1^{k_s}, \dots, f_n^{k_s})$, $s = 1, 2, \dots$ and then the corresponding sequence $f^{k_s} = f_1^{k_s} e_1 + \dots + f_n^{k_s} e_n$ is fundamental in H since

$$\|f^{k_s} - f^{k_p}\|^2 = \sum_{i=1}^{n} |f_i^{k_s} - f_i^{k_p}|^2 \to 0$$

for $s, p \to \infty$. The necessity was proven in Section 2 of Chapter 1. ■

Let H be a finite-dimensional separable Hilbert space and let $e_1, e_2, \dots, e_n, \dots$ be its orthonormed basis. Denote by P_n' the orthoprojector mapping H onto the n-dimensional space L_n formed by the elements $e_1, \dots, e_n$ and let $P_n'' = I - P_n'$. Then for $\forall\, f \in H$ and $\forall\, n \ge 1$ we have (see Section 8 of Chapter 1):

$$f = P_n' f + P_n'' f, \tag{13.1}$$

where $P_n' f = \sum_{k=1}^{n} f_k e_k$, $P_n'' f = \sum_{k=n+1}^{\infty} f_k e_k$, $f_k = (f, e_k)$, and

$$\|f\|^2 = \|P_n' f\|^2 + \|P_n'' f\|^2. \tag{13.2}$$

More than that,

$$\|P_n' f\|^2 = \sum_{k=1}^{n} |f_k|^2, \quad \|P_n'' f\|^2 = \sum_{k=n+1}^{\infty} |f_k|^2. \tag{13.3}$$

Consequently, for $\forall\, f \in H$ the number sequence $\|P_n'' f\|^2$, $n = 1, 2, \dots$ tends monotonically to zero for $n \to \infty$. Let us formulate a theorem showing that if a set $M \subset H$ is bounded and "almost finite-dimensional," it is necessary and sufficient for it to be compact.

Theorem 1. *If a set $M \subset H$ is bounded and for any $\varepsilon > 0$ there exists a number $n = n(\varepsilon)$ such that $\|P_n''f\| \leq \varepsilon$ for all $f \in M$, it is necessary and sufficient for it to be compact.*

□ *Sufficiency.* Since there exists $C > 0$ such that $\|f\| < C$ for $\forall f \in M$, then by (13.2) $\|P_n'f\| < C$ too; that is, there is a set of elements $P_n'f$, where $f \in M$ is bounded. It is a finite-dimensional set and, by Lemma 1, compact. Consequently, by the Hausdorff theorem there exists in it for $\forall \varepsilon > 0$ a finite ε-net. By the condition $\|P_n''f\| < \varepsilon$ for $\forall f \in M$, elements of this ε-net form 2ε-net for all the set M; therefore, M is compact.

Necessity. Let the set M be compact. Then there exists for it a finite ε-net for $\forall \varepsilon > 0$. Let the elements $\xi_1, \xi_2, \ldots, \xi_{N_\varepsilon}$ be elements of this ε-net. Let us take among them just linearly independent and form with the Sonin–Schmidt orthogonalization process, the orthonormed system $\bar{e}_1, \bar{e}_2, \ldots, \bar{e}_{\bar{N}_\varepsilon}$ where $\bar{N}_\varepsilon \leq N_\varepsilon$. Then $\forall f \in M$ is representable in the form

$$f = \sum_{i=1}^{\bar{N}_\varepsilon} \bar{f}_i \bar{e}_i + \eta, \quad \|\eta\| < \varepsilon$$

and $|\bar{f}_i| < C$; consequently,

$$P_n''f = \sum_{i=1}^{\bar{N}_\varepsilon} \bar{f}_i P_n'' \bar{e}_i + P_n''\eta \tag{13.4}$$

and

$$\|P_n''f\| \leq C\bar{N}_\varepsilon \max_i \|P_n''\bar{e}_i\| + \varepsilon.$$

Since there is a finite number of elements $\bar{e}_i$, the number n_0 will be found such that for $n > n_0$ $\max_i \|P_n''\bar{e}_i\| < \varepsilon/(C\bar{N}_\varepsilon)$, and then

$$\|P_n''f\| < 2\varepsilon \quad \text{for} \quad \forall f \in M. \blacksquare$$

2. Compact operators and their properties. Theorem on the representation of compact operator

Linear operator $A \in \mathcal{L}(\mathbf{X}, \mathbf{Y})$ is referred to as *compact* (or *completely continuous*) if it transforms any bounded in $\mathbf{X}$ set into a compact in $\mathbf{Y}$ set. The linear functional $\langle f, x\rangle \in \mathcal{L}(\mathbf{X}, \mathbf{R}^1)$ is an example.

Let $\mathbf{X} = \mathbf{Y} = H$. It is easy to show that if the operators $A_1, A_2 \in \mathcal{L}(H)$ are compact, then for any scalar λ, μ an operator $\lambda A_1 + \mu A_2$ is compact, and if an operator $A \in \mathcal{L}(H)$ is compact and an operator $B \in \mathcal{L}(H)$, then the operators AB and BA are compact. It follows by the lemma on the boundedness of a compact set that the compact operator is bounded. However, not every bounded operator is compact. For example, the unit operator I in finite-dimensional space H cannot be compact, since it transforms a noncompact bounded set, noncompact orthonormed basis, into itself. The operator B inverse to a compact operator is obligatorily unbounded, since otherwise the operator $AB = I$ would be compact. The finite-dimensional operator and adjoint to it operator are compact. It follows from Lemma 1.

The following lemma results from the statement that if a set M is compact, then the set $M' = \{\alpha x : x \in M\}$ is compact too for any fixed scalar α.

Lemma 2. *If an operator $A \in \mathcal{L}(H)$ transforms the unit closed ball into a compact set, then A is a compact operator.*

There is an important theorem that often helps to justify the substitution of an equation containing a compact operator by an equation with a finite-dimensional operator (8.5) approximate to it.

Theorem 2. *It is necessary and sufficient for a linear bounded operator $A \in \mathcal{L}(H)$ assigned in separable Hilbert space H to be compact, that for any $\varepsilon > 0$ integer number $n = n(\varepsilon)$ and linear operator A_1 and A_2 (A_1 is n-dimensional and $\|A_2\| \leq \varepsilon$) will be found such that*

$$A = A_1 + A_2. \tag{13.5}$$

All compact operators are thus "almost finite-dimensional" operators.

□ *Necessity.* By (13.1) for $\forall f \in H$ and $\forall n > 0$ the expansion exists:

$$Af = P'_n Af + P''_n Af \qquad (A = P'_n A + P''_n A).$$

Since the compactness of the set $\{Af/\|f\|\}$ follows from the boundedness of the set $\{f/\|f\|\}$, then by Theorem 1 there exists $n_0 = n_0(\varepsilon)$ for $\forall \varepsilon > 0$ such that $\|P''_n Af\| = \|f\| \|P''_n Af/\|f\|\| \leq \varepsilon \|f\|$ for $n > n_0(\varepsilon)$. The operator $P'_n A$ is n-dimensional since $P'_n Af = \sum\limits_{k=1}^{n} (Af, e_k) e_k = \sum\limits_{k=1}^{n} (f, A^* e_k) e_k$; consequently, the necessity

$P'_n Af = \sum_{k=1}^{n} (Af, e_k)e_k = \sum_{k=1}^{n} (f, A^* e_k)e_k$; consequently, the necessity is proven.

Sufficiency. It is enough to show that operator A, if there exists expansion (13.5) for it, transforms the unit closed ball $\bar{D}_1(0)$ into the compact set $\bar{R}(A)$. In order to prove the compactness of $\bar{R}(A)$ let us show that for the set $\bar{R}(A)$ there always exists for $\forall\varepsilon > 0$ a finite 2ε-net.

Indeed, a set of elements of type $A_1 x$ where $x \in \bar{D}_1(0)$ is always compact because A_1 is finite-dimensional; therefore, there exists in it a finite ε-net M_ε. Under condition $\|A_2\| < \varepsilon$ the set M_ε is 2ε-net for all the set $\bar{R}(A)$. ■

Corollary 1. If $A \in \mathcal{L}(H)$ is a compact operator, then adjoint to it operator A^* is compact too.

□ In fact, by (13.5) [see (8.5′)] $A^* = A_1^* + A_2^*$ where $\|A_2^*\| \leq \varepsilon$ and A_1^* is a finite-dimensional operator. ■

3. Equation of the second kind with compact operator. Formulation of the Fredholm theorems

Take in H an equation called the *equation of the second kind*

$$u = Au + f, \tag{13.6}$$

where $u, f \in H$, the operator A is compact, and the equation

$$v = A^* v + g, \tag{13.7}$$

where $v, g \in H$, is *its adjoint equation.* It proves to be possible to transform Equations (13.6) and (13.7) with compact operators into equivalent equations with finite-dimensional operators that, in turn, are equivalent to some self-conjugate systems of linear algebraic equations. This fact permits for Equations (13.6) and (13.7) the formulation of famous Fredholm theorems on the basis of corresponding theorems of linear algebra on the existence and uniqueness of solutions to systems of linear algebraic equations.

Let us show how Equations (13.6) and (13.7) are transformed into equations with finite-dimensional operators. It is possible, by Theorem 2, to rewrite Equation (13.6) in the form $(I - A_2)u = A_1 u + f$, where A_1 is an n-dimensional operator and $\|A_2\| \leq \varepsilon < 1$.

exists, and Equation (13.6) can be rewritten as

$$h = A_1(I - A_2)^{-1}h + f. \tag{13.8}$$

Since $\|A_2^*\| \leq \varepsilon < 1$, there exists the bounded operator $(I - A_2^*)^{-1} = [(I - A_2)^{-1}]^*$; therefore, rewriting Equation (13.7) as $(I - A_2^*)v = A_1^*v + g$, and multiplying it from the left by the operator $(I - A_2^*)^{-1}$, we arrive at the equivalent equation

$$v = [(I - A_2^{-1})]^* A_1^* v + z, \tag{13.9}$$

where $z = (I - A_2)^{-1}g$ and the operator $[(I - A_2^{-1})]^* A_1^*$ is adjoint to the operator $A_1(I - A_2)^{-1}$ [in Equation (13.8)].

The operator $A_1(I - A_2)^{-1}$ and, consequently, its adjoint operator, are n-dimensional operators. If we denote by $e_1, e_2, \ldots, e_n, \ldots$ an orthonormed basis, where $e_1, e_2, \ldots, e_n$ belong to the domain of values of the operator $A_1(I - A_2)^{-1}$, then calculations show that Equation (13.8) is equivalent to a system of algebraic equations of an order n for the determination of Fourier coefficients $h_1, h_2, \ldots, h_n$ of desired element h (coefficients h_j for $j > n$ are determined easily). Similarly, Equation (13.9) is equivalent to a system of algebraic equations of an order n for the determination of Fourier coefficients v_j of an element v $(j \leq n)$.

It is obvious, that preceding is just and applicable to the following equations

$$u = \mu A u + f \tag{13.10}$$

$$v = \bar{\mu} A^* v + g, \tag{13.11}$$

where μ is the numerical, generally speaking, complex parameter. The preceding reasoning implies the following statements known as the Fredholm theorems.

Fredholm theorems. If μ^{-1} is not an eigenvalue of an operator A, it is necessary and sufficient for Equation (13.10) *to have a unique solution for all $f \in H$.*

If μ^{-1} is the eigenvalue, then its multiplicity is finite and $\bar{\mu}^{-1}$ is an eigenvalue of the operator A^* of the same multiplicity.

If an element f is orthogonal to all eigenelements of the operator A^* corresponding to the eigenvalue $\bar{\mu}^{-1}$, it is necessary and sufficient for Equation (13.10) *to be resolvable.*

More than that, there exists a unique solution to Equation (13.10) *orthogonal to all eigenelements of the operator* A^*, *corresponding to the eigenvalue* $\bar{\mu}^{-1}$.

We cite these theorems here without proofs; however, diligent and industrious readers, following the guidelines of our reasoning preceding these theorems, can, using theorems of linear algebra, prove them themselves. Those who are familiar with the theory of integral equations can almost literally apply to the case under consideration corresponding methods for the proof of integral Fredholm theorems.

4. Eigenvalues and eigenelements of compact operator

Lemma 3. *If an operator* $A \in \mathcal{L}(H)$ *is compact,*[1] *then eigensubspace* H_λ *corresponding to an eigenvalue* $\lambda \neq 0$ *is finite-dimensional.*

□ Let the subspace $H_\lambda \subset H$ be infinite-dimensional and let $\{e_n\}_1^\infty$ be an orthonormed system of elements in H_λ. Then $e_n = (1/\lambda)Ae_n$ and, consequently, the sequence $\{e_n\}_1^\infty$ is compact, which is impossible, since $\|e_n - e_k\|^2 = 2$ for $n \neq k$. ■

Theorem 4 (*the fourth Fredholm theorem*). *Only a finite number of eigenvalues of a compact operator* $A \in \mathcal{L}(H)$ *can be contained for any* $M > 0$ *outside a circle* $|\lambda| < M$ *of the complex plane.*

□ Assume, on the contrary, that there is outside the circle $|\lambda| < M$ an infinite number of eigenvalues $\lambda_1, \lambda_2, \ldots, \lambda_n, \ldots$ that can be chosen being different by Lemma 3 ($\lambda_i \neq \lambda_j$, $i \neq j$). Let e_i be any eigenelement corresponding to an eigenvalue λ_i. Show that for $\forall n \geq 1$ the system $\{e_k\}_1^n$ is linearly independent. This statement is obvious for $n = 1$. Let it be valid for $n - 1$. If one assumes that $\{e_k\}_1^n$ is linearly dependent, then $e_n = c_1e_1 + \ldots + c_{n-1}e_{n-1}$ for some constants $\{c_k\}_1^{n-1}$ not all equal to zero. But, then $Ae_n = \lambda_n e_n = \sum_{k=1}^{n-1} c_k\lambda_k e_k$ whence $\sum_{k=1}^{n-1} c_k (1 - \lambda_k/\lambda_n) e_k = 0$ which is impossible, since $1 - \lambda_k/\lambda_n \neq 0$, $k = 1, \ldots, n-1$.

Denote by H_n a subspace formed by the elements $\{e_k\}_1^n$. It follows from what we had proven, that $H_1 \subset H_2 \subset \ldots \subset H_n \subset \ldots$,

[1] Here H is a complex space.

$H_n \neq H_{n-1}$ for $n > 1$. Therefore, an element $f_n \in H_n$, $f_n \perp H_{n-1}$, $\|f_n\| = 1$ will be found for $\forall\, n > 1$. Then it is possible to separate the fundamental sequence from the set $\{Af_k\}_1^\infty$. In fact, it is impossible, which is the contradiction proving the theorem. Then we have for arbitrary m and n, $m < n$,

$$Af_n - Af_m = \lambda_n f_n + \lambda_n \left(\frac{1}{\lambda_n} Af_n - f_n \right) - Af_m = \lambda_n f + \sigma_n,$$

where $\sigma_n \in H_{n-1}$, since $Af_m \in H_m \subset H_{n-1}$ and

$$\frac{1}{\lambda_n} A_n f_n - f_n = \frac{1}{\lambda_n} A \left(\sum_{k=1}^{n} c_k e_k \right) - \sum_{k=1}^{n-1} c_k e_k$$

$$= \sum_{k=1}^{n-1} c_k \left(\frac{\lambda_k}{\lambda_n} - 1 \right) e_k \in H_{n-1}.$$

Therefore,

$$\|Af_n - Af_m\|^2 = \|\lambda_n f_n\|^2 + \|\sigma_n\|^2 \geq (|\lambda_n|\, \|f_n\|)^2 = |\lambda_n|^2 \geq M^2. \blacksquare$$

Theorem 5. *If an operator $A \in \mathcal{L}(H)$ is selfadjoint and compact, then the number $M = \sup\limits_{\|u\|=1} (Au, u)$ [similarly, the number $m = \inf\limits_{\|u\|=1} (Au, u)$], if it differs from zero, is an eigenvalue of this operator.*

□ Let $M \neq 0$, then we have for $\forall\, u \in H$: $(Mu - Au, u) \geq 0$. Show that there exists $u_0 \in H$, $u_0 \neq 0$ such that $(Mu_0 - Au_0, u_0) = 0$. Then the statement of the theorem is valid by virtue of the Theorem 2, Section 9 of this chapter.

The supposition that such element u_0 does not exist would mean that $(Mu - Au, u)$, $u \in H$ may turn to zero just for $u = 0$. Therefore, bilinear form $(Mu - Au, v)$ could be taken as the new inner product in H for which the generalized Cauchy–Bunyakovskii inequality (8.8) would be valid

$$|(Mu - Au, v)|^2 \leq (Mu - Au, u)(Mv - Av, v). \tag{13.12}$$

There exists, by the definition of M, a sequence $u_1, u_2, \ldots$ $\|u_i\| = 1$, $i = 1, 2, \ldots$ for which $(Au_n, u_n) \to M$ or

$$(Mu_n - Au_n, u_n) \to 0 \quad \text{for} \quad n \to \infty. \tag{13.13}$$

Assign $u = u_n$, $v = Mu_n - Au_n$ in (3.13) to obtain

$$\|Mu_n - Au_n\|^4 \le (Mu_n - Au_n, u_n)((MI - A)^2 u_n, (MI - A)u_n)$$

$$\le (Mu_n - Au_n, u_n)(|M| + \|A\|)^3. \quad (13.14)$$

It follows from (13.14) that $Mu_n - Au_n \to 0$ for $n \to \infty$. But the sequence Au_n is compact, which means that it is possible to separate from it the converging subsequence Au_{n_k}; then the sequences Mu_{n_k} and u_{n_k} also converge to some limits Mu_0 and u_0. Obviously, $\|u_0\| = 1$ and by (13.14), $(Mu_0 - Au_0, u_0) = 0$. The contradiction proves the theorem. ■

§ 14. The Sobolev Spaces. Embedding Theorems

General definition of the space $W_p^l(\Omega)$. Hilbert spaces $W_2^l(\Omega)$, $\overset{\circ}{W}{}_2^l(\Omega)$, $W_2^1(\Omega)$, $\overset{\circ}{W}{}_2^1(\Omega)$. Generalized derivatives. Theorem on traces. Poincaré–Steklov inequality. Embedding operator. Formulation of embedding theorems. Simple embedding theorem.

Relationships between differential properties of functions play an important role in various fields of analysis and mathematical physics. They are of fundamental significance in applied mathematical physics and numerical mathematics and served as a basis for the creation of a new theory, the *theory of generalized solutions* of boundary value problems for partial differential equations. S. Sobolev created this theory.

1. General definition of the space $W_p^l(\Omega)$. Hilbert spaces $W_2^l(\Omega)$, $\overset{\circ}{W}{}_2^l(\Omega)$, $W_2^1(\Omega)$, $\overset{\circ}{W}{}_2^1(\Omega)$. Generalized derivatives

Let a bounded domain Ω with sufficiently smooth boundary $\partial\Omega$ (for instance, satisfying the Lipschitz condition) be a set in $\mathbf{R}^n$. In the Sections 1 and 6 of Chapter 1 we introduced the space $\tilde{L}_p(\Omega)$ and its completion $L_p(\Omega)$, the Lebesgue space, having defined the norms for elements of this space with formula (6.17), Chapter 1 where the integral is understood in the Lebesgue sense [see (1.22), Chapter 1]. Let us generalize the performed constructions as follows.

Let $C^l(\bar{\Omega})$ be a class of functions $u(x)$, l times continuously differentiable on $\bar{\Omega}$, $p \geq 1$. Take the norm on this space

$$\|u\|_{W_p^l} = \left[\int_\Omega \left(|u|^p + \sum_{1\leq|\alpha|\leq l} |D^\alpha u|^p\right) d\Omega\right]^{1/p}, \tag{14.1}$$

where D^α means the partial derivative of an order $|\alpha|$, $\alpha = (\alpha_1, \alpha_2, \ldots, \alpha_n)$ is a multiindex, and $|\alpha| = \sum \alpha_i$. Denote normed space so obtained by $\tilde{W}_p^l(\Omega)$. Denote its completion with respect to norm (14.1) by $W_p^l(\Omega)$. This Banach space is called the *Sobolev space.*

The case $p = 2$ is important; then we sometimes denote the space $W_2^l(\Omega)$ by $H^l(\Omega)$. The space $W_2^l(\Omega)$ is a Hilbert space-completion of $\tilde{W}_2^l(\Omega)$ with respect to the norm generated by the inner product

$$[u, v]_{W_2^l} = \int_\Omega \left(uv + \sum_{1\leq|\alpha|\leq l} D^\alpha u D^\alpha v\right) d\Omega. \tag{14.2}$$

This norm looks like

$$\|u\|_{W_2^l} = \left(\|u\|_{L_2}^2 + \sum_{1\leq|\alpha|\leq l} \|D^\alpha u\|_{L_2}^2\right)^{1/2}. \tag{14.3}$$

It is possible to write down the equivalent norm instead of (14.3):

$$\|u\|_{W_2^l} = \|u\|_{L_2} + \sum_{1\leq|\alpha|\leq l} \|D^\alpha u\|_{L_2}. \tag{14.4}$$

Inner product (14.2) for the space $W_2^1(\Omega)$ takes the form

$$[u, v]_{W_2^1} = \int_\Omega \left(\sum_{i=1}^n \frac{\partial u}{\partial x_i}\frac{\partial v}{\partial x_i} + uv\right) d\Omega \tag{14.5}$$

and the norm

$$\|u\|_{W_2^1} = \left[\int_\Omega \left(\sum_{i=1}^n \left(\frac{\partial u}{\partial x_i}\right)^2 + u^2\right) d\Omega\right]^{1/2}. \tag{14.6}$$

It follows from (14.3) that if $u \in W_2^l(\Omega)$, then $u \in L_2(\Omega)$ and $D^\alpha u \in \in L_2(\Omega)$ for $1 \leq |\alpha| \leq l$. Now, what should be understood as a derivative of a function u? Let us consider this question in more detail.

First, introduce two new spaces of functions. Denote by $C_0^l(\bar{\Omega})$ the normed space l times continuously differentiable in $\bar{\Omega}$ functions, with the norm of the space $C^l(\overline{\Omega})$, that turns to zero in some neighborhood (different for different functions) of the boundary $\partial\Omega$. Let $C^{l,\alpha}(\bar{\Omega})$ denote the space for all functions from $C^l(\bar{\Omega})$, whose lth derivatives satisfy the Hölder condition with the index $\alpha(\alpha \in (0,1))$ and with the norm

$$\|u\|_{C^{l,\alpha}} = \|u\|_{C^l} + \max_{|\beta|=l} \sup_{\substack{x,y\in\Omega \\ x\neq y}} \frac{|D^\beta u(x) - D^\beta u(y)|}{\|x-y\|^\alpha}, \qquad (14.7)$$

where $\|\cdot\|$ means the Euclidean norm in $\mathbf{R}^n$ [see (6.4), Chapter 1).

Let $u \in C^l(\bar{\Omega})$ and let φ be any function from $C_0^l(\bar{\Omega})$. Apply k times $(1 \leq k \leq l)$ the formula of integrating by parts to obtain

$$\int_\Omega D^k u\varphi \, d\Omega = (-1)^{|k|} \int_\Omega u D^k \varphi \, d\Omega. \qquad (14.8)$$

We use equality (14.8) for the definition of *generalized derivatives.* We say that a function $v = D^k u \in L_2(\Omega)$ is called a *generalized derivative of an order k* of a function $u \in L_2(\Omega)$ if for any function $\varphi \in C_0^l(\bar{\Omega})$ equality (14.8) holds. The generalized derivative is determined uniquely. Indeed, if together with v a function $\tilde{v}$ would satisfy for $\forall\, \varphi \in C_0^l(\bar{\Omega})$ Equation (14.8):

$$\int_\Omega \tilde{v}\varphi \, d\Omega = (-1)^{|k|} \int_\Omega u D^k \varphi \, d\Omega \qquad (14.9)$$

then, subtracting (14.9) from (14.8) we would get

$$\int_\Omega (v - \tilde{v})\varphi \, dx = 0, \qquad \forall\, \varphi \in C_0^l(\bar{\Omega}).$$

Since the space $C_0^l(\bar{\Omega})$ is everywhere dense in $L_2(\Omega)$, then $v = \tilde{v}$ almost everywhere, but we do not distinguish such functions as elements of $L_2(\Omega)$. Now let $\{u_n(x)\}$ be fundamental in the $W_2^l(\Omega)$ sequence of functions from $C^l(\bar{\Omega})$ that determines the function $u \in W_2^l(\Omega)$. Equality (14.8) holds for all $u_n(x)$; passing to a limit for $n \to \infty$, we prove it for the limit function too, and prove that the function $u \in W_2^l \Omega$ has all generalized derivatives $D^k u$, $1 \leq k \leq l$.

2. Formulation of the theorem on traces of functions from the space $W_2^1(\Omega)$. The space $\overset{\circ}{W}{}_2^l(\Omega)$. The Poincaré–Steklov inequality

Values $u(s)$, $s \in \partial\Omega$ on the boundary $\partial\Omega$ of each function $u \in C^l(\Omega)$ are identically determined; the function $u(s)$ is called a *trace of the function* $u(x)$. Obviously, it is continuous on $\partial\Omega$ and, consequently, integrable on $\partial\Omega$ with a square ($u(s) \in L_2(\partial\Omega)$). The following theorem, which we present without proof, permits the extension of the notion of trace to all functions $u \in W_2^l(\Omega)$, $l \geq 1$ that in certain cases have no continuous extension to a boundary.

Theorem 1. *Let Ω be a domain with a boundary that satisfies the Lipschitz condition. Then there is a unique bounded linear operator $T \in \mathcal{L}(W_2^1(\Omega), L_2)$ such that $Tu(x) = u(s)$ if $u \in C^1(\bar{\Omega})$.*

By this theorem, each function $u \in W_2^1(\Omega)$ has a corresponding function $u(s) \in L_2(\partial\Omega)$, which is defined as its trace on the boundary $\partial\Omega$. This correspondence, due to the boundedness of T, is such that each pair of close in $W_2^1(\Omega)$ functions has corresponding on $\partial\Omega$ a pair of close in $L_2(\partial\Omega)$ functions. Since the set $C^1(\bar{\Omega})$ is everywhere dense in $W_2^1(\Omega)$ the trace $u(s)$ of each function $u \in W_2^1(\Omega)$, $u \notin C^1(\Omega)$ can be regarded as a limit in $L_2(\partial\Omega)$ of a sequence of traces $\{u_n(s)\}$ of functions $u_n(x) \in C^1(\bar{\Omega})$ that converge in $W_2^1(\Omega)$ to the function u. It is obvious that if $u \in W_2^1(\Omega) \cap C(\bar{\Omega})$, then its trace $u(s)$ is assigned by its values on a boundary. One can show that a space of traces of functions $u \in W_2^1(\Omega)$ denoted by $W_2^{1/2}(\partial\Omega)$ is everywhere dense in $L_2(\partial\Omega)$. However, not every function on $L_2(\partial\Omega)$ belongs to $W_2^{1/2}(\partial\Omega)$. Since for $\forall\, u \in W_2^l(\Omega)$ the functions $D^\alpha u \in W_2^1(\Omega)$ for $|\alpha| < l$, then it is possible to speak about the traces of the function $D^\alpha u$ on $\partial\Omega$.

We have defined the trace of a function u on a manifold of dimension $n-1$. One may generalize the preceding approach and speak about the traces of these functions as elements of some spaces L_q on the manifolds of lower dimension.

Let us define the inner product (14.2) on the functions of the class $C_0^l(\bar{\Omega})$ and complete this space with respect to norm (14.3). As a result, we obtain new Hilbert space that we denote by $\overset{\circ}{W}{}_2^l(\Omega)$.

By the theorem on traces, all functions of this space together with generalized derivatives up to the order $l-1$ have zero traces on $\partial\Omega$.

Let us examine the space $\overset{\circ}{W}{}_2^1(\Omega)$ in more detail. Prove that for the functions $u \in \overset{\circ}{W}{}_2^1(\Omega)$ the so-called *Poincaré–Steklov inequality*

$$\int\limits_\Omega u^2 d\Omega \le C \int\limits_\Omega \sum_{i=1}^n \left(\frac{\partial u}{\partial x_i}\right)^2 d\Omega \tag{14.10}$$

holds where $C > 0$ is a constant determined only by the characteristics of the domain Ω.

□ Choose a system of coordinates in $\mathbf{R}^n$ so that the domain Ω will be inside some parallelepiped $P\{a_k \le x_k \le b_k,\ k = \overline{1,n}\}$. Let $l_k = b_k - a_k$. It is sufficient to prove the inequality for the function $u_n \in C_0^1(\overline{\Omega})$ since then, passing to a limit in equality (14.10) for $n \to \infty$ written for the functions u_n, that tend to $u \in \overset{\circ}{W}{}_2^1(\Omega)$, we obtain this inequality for the whole class of functions $\overset{\circ}{W}{}_2^1(\Omega)$.

So, let $u \in C_0^1(\bar{\Omega})$. Complete a definition of the function $u(x)$ by zero in the domain $P\backslash\Omega$. It will be continuous in the $\bar{P}$ function $u|_{\partial\Omega} = 0$ and its derivatives can have a discontinuity while passing through the boundary $\partial\Omega$. The Newton–Leibniz formula is valid for such functions. Let a point $x = (x_1, x_2, \ldots, x_n) \in P$; denote it as $x = (x_1, x')$ where $x' = (x_2, \ldots, x_n)$. Then

$$u(x) - u(a_1, x') = \int_{a_1}^{x_1} \frac{\partial u(\xi, x')}{\partial \xi} d\xi$$

but $u(a_1, x') = 0$, therefore, by the Cauchy–Bunyakovskii inequality,

$$u^2(x) \le \int\limits_{a_1}^{x_1} d\xi \int\limits_{a_1}^{x_1} \left(\frac{\partial u(\xi, x')}{\partial \xi}\right)^2 d\xi$$

$$= (x_1 - a_1) \int_{a_1}^{x_1} \left(\frac{\partial u(\xi, x')}{\partial \xi}\right)^2 d\xi. \tag{14.11}$$

Similarly,

$$u^2(x) \le (b_1 - x_1) \int\limits_{x_1}^{b_1} \left(\frac{\partial u(\xi, x')}{\partial \xi}\right)^2 dx. \tag{14.12}$$

Divide (14.11) and (14.12), respectively, by $(x_1 - a_1)$ and $(b_1 - x_1)$, sum and, with regard to

$$\frac{1}{x_1 - a_1} + \frac{1}{b_1 - x_1} = \frac{l_1}{(x_1 - a_1)(b_1 - x_1)},$$

obtain

$$u^2(x) \leq \frac{(x_1 - a_1)(b_1 - x_1)}{l_1} \int_{a_1}^{b_1} \left(\frac{\partial u(\xi, x')}{\partial \xi} \right)^2 d\xi. \tag{14.13}$$

Let us integrate this inequality with respect to the parallelepiped $P' = \{a_k \leq x_k \leq b_k,\, k = \overline{2, n}\}$:

$$\int_{P'} u^2 dx' \leq \frac{(x_1 - a_1)(b_1 - x_1)}{l_1} \int_P \left(\frac{\partial u}{\partial x_1} \right)^2 d\Omega. \tag{14.14}$$

Integrate (14.14) once more with respect to x_1 from a_1 to b_1

$$\int_P u^2 d\Omega \leq \frac{(l_1^2)}{6} \int_P \left(\frac{\partial u}{\partial x_1} \right)^2 d\Omega.$$

Discard integrals with respect to $P \backslash \Omega$ on the left and on the right that are equal to zero. Perform the same laying out with each variable x_k to obtain the inequalities

$$\frac{6}{l_k^2} \int_\Omega u^2\, d\Omega \leq \int_\Omega \left(\frac{\partial u}{\partial x_k} \right)^2 d\Omega, \qquad k = 1, \ldots, n.$$

Summing up these inequalities we obtain inequality (14.10) with a constant

$$C = (\sigma \sum_{k=1}^n l_k^{-2})^{-1}. \quad \blacksquare$$

We have arrived at important corollary.

Corollary 1. *A seminorm*

$$\|u\|_{\overset{\circ}{W}{}_2^1} = \left(\int_\Omega \sum_{i=1}^n \left(\frac{\partial u}{\partial x_i} \right)^2 d\Omega \right)^{1/2} \tag{14.15}$$

is a norm in the space $\overset{\circ}{W}{}_2^1(\Omega)$. *Norms* (14.15) *and* (14.6) *are equivalent.*

Lemma 2. *Let $A > 0$ be a symmetric operator; then the generalized Cauchy–Bunyakovskii inequality is valid for $\forall\, x, y \in D(A)$:*

$$|(Ax, y)| \leq (Ax, x)^{1/2}(Ay, y)^{1/2}. \tag{8.8}$$

□ Define the new inner product on $D(A)$ with the formula

$$[x, y] = (Ax, y). \tag{8.9}$$

It satisfies all axioms of Hermitian space and, consequently, the Cauchy–Bunyakovskii inequality

$$|[x, y]| \leq [x, x]^{1/2}[y, y]^{1/2}$$

holds for it, which is inequality (8.8). ■

We had introduced in Section 2 of this chapter the operator P of orthogonal projection from a space H onto some of its subspace L and investigated some of its properties; let us continue this study.

Lemma 3. *It is necessary and sufficient for linear operator P in H to be an orthoprojector, that:*

(1) *P should be a selfadjoint operator;*

(2) $P^2 = P$.

□ *Necessity.* If $x_1 = y_1 + z_1$, $x_2 = y_2 + z_2$, where $y_1, y_2 \in L$ and $z_1, z_2 \in L^{\perp}$, then

$$(Px_1, x_2) = (y_1, y_2 + z_2) = (y_1, y_2) + (y_1, z_2)$$

$$= (y_1, y_2) = (y_1 + z_1, y_2) = (x_1, Px_2).$$

The self-adjointness is proven; finally, $P^2 x_1 = P(Px_1) = Py_1 = Px_1$.

Sufficiency. Denote by L a set of all $x \in H$ for which $Px = x$; obviously, L is a subspace in H. Let x be an arbitrary element from H, then

$$x = Px + (x - Px) \tag{8.10}$$

and $Px \in L$, since $P(Px) = Px$. If $u \in L$, then $Pu = u$. Consequently,

$$(x - Px, u) = (x, u) - (Px, u) = (x, u) - (x, Pu) = 0;$$

(1) $W_p^l(\Omega) \hookrightarrow L_{p^*}(\Omega)$ *for* $1/p^* = 1/p - l/n$ *if* $l < n/p$;

(2) $W_p^l(\Omega) \hookrightarrow L_q(\Omega)$ *for all* $q \in [1, \infty)$ *if* $l = n/p$;

(3) $W_p^l(\Omega) \hookrightarrow C^{0,l-n/p}(\overline{\Omega})$ *if* $n/p < l < n/p + 1$;

(4) $W_p^l(\Omega) \hookrightarrow C^{0,\alpha}(\overline{\Omega})$ *for* $0 < \alpha < 1$ *if* $l = n/p + 1$;

(5) $W_p^l(\Omega) \hookrightarrow C^{0,1}(\bar{\Omega})$ *if* $n/p + 1 < l$;

(6) *if* $u \in W_p^l(\Omega)$, $n > lp$, *then* $u \in L_{q^*}$ *too on any intersection of* Ω *with a hyperplane of* s *dimensions where* $s > n - lp$, $q^* < sp/(n - lp)$.

By the Sobolev–Kartashov theorem there are the embeddings:

(7) $W_p^l(\Omega) \overset{c}{\hookrightarrow} L_q(\Omega)$ *for all* $1 \le q < p^*$ *for* $1/p^* = 1/p - l/n$ *if* $l < n/p$;

(8) $W_p^l(\Omega) \overset{c}{\hookrightarrow} L_q(\Omega)$ *for* $\forall\, q \in [1, \infty)$ *if* $l = n/p$;

(9) $W_p^l(\Omega) \overset{c}{\hookrightarrow} C(\overline{\Omega})$ *if* $l > n/p$.

Corollary. *If* $n < (l - s)p$, *then* $W_p^l(\Omega) \overset{c}{\hookrightarrow} C^s(\bar{\Omega})$.

Proof of the preceding theorems would require the application of the cumbersome apparatus of the theory of functions, refining assumptions on the properties of a domain and its boundary, and so forth. We restrict ourselves to the proof of a simple embedding theorem.

Theorem 3. *The space* $W_p^1(a, b)$ *for* $p > 1$ *is embedded into* $C^{0,q^{-1}}[a, b]$ *if* $p^{-1} + q^{-1} = 1$.

□ First, let $u(x) \in C^1[a, b]$. According to the theorem on average $\exists \xi \in [a, b]$ such that $u(\xi) = (b - a)^{-1} \int_a^b u(s)ds$. Therefore, the following equality is valid on $[a, b]$:

$$u(x) = \frac{1}{b-a}\int_a^b u(s)ds + \int_\xi^x u'(s)ds.$$

Applying the Hölder theorem [(1.24), Chapter 1] with unit multiplier we have

$$|u(x)| \le (b-a)^{1/q}\left(\int_a^b |u'(s)|^p\, ds\right)^{1/p}$$

$$+ (b-a)^{-1/p}\left(\int_a^b |u(s)|^p\, ds\right)^{1/p} \le C\|u\|_{W_p^1(a,b)},$$

where $C = \max(b-a)^{1/q}, (b-a)^{-1/p}$ which means that

$$\|u\|_{C[a,b]} \le C\|u\|_{W_p^1(a,b)}. \tag{14.16}$$

Now let a sequence $\{u_n(x)\} \in C^1[a,b]$ be fundamental in $W_p^1(a,b)$. Then use inequality (14.16) to obtain

$$\|u_n - u_m\|_{C[a,b]} \le C\|u_n - u_m\|_{W_p^1(a,b)} \to 0$$

for $n, m \to \infty$; consequently, $\{u_n\}$ is fundamental in $C[a,b]$ too, that is, it converges obligatory to continuous function $u(x)$. Therefore, the elements $W_p^1(a,b)$ can be identified with continuous functions. Let $u_n(x) \to u(x)$ in $W_p^1(a,b)$. Pass in inequality (14.16) for $u = u_n$ to a limit for $n \to \infty$ to convince ourselves that it is valid for $\forall\, u \in W_p^1(a,b)$. So, for $p \ge 1$ $W_p^1(a,b) \subset C[a,b]$ and it is possible to speak about a value of $u(x)$ in any point $x_0 \in [a,b]$. Let $p > 1$ and let us estimate the second term in formula (14.7) for a norm in the space $C^{0,1/q}[a,b]$ that in this case has the form

$$\beta(u) = \max_{\substack{x,y\in[a,b]\\ x\neq y}} \frac{|u(x)-u(y)|}{|x-y|^{1/q}}.$$

Again, let first $u(x) \in C^1[a,b]$, then we have for $x \ne y$

$$|u(x) - u(y)| = \left|\int_y^x u'(s)\, ds\right| \le |x-y|^{1/q}\left(\int_a^b |u'|^p\, ds\right)^{1/p};$$

that is,

$$\beta(u) \le \|u\|_{W_p^1(a,b)}.$$

Consequently, with regard to (14.16)

$$\|u\|_{C^{0,1/q}[a,b]} \le (C+1)\|u\|_{W_p^1(a,b)}. \tag{14.17}$$

Repeating the reasoning related to inequality (14.16), we obtain that (14.17) is valid for $\forall\, u(x) \in W_p^1(a,b)$; that is, $W_p^1(a,b) \subset C^{0,1/q}[a,b]$ for $p > 1$. ■

Corollary 1. *For* $p > 1$ $W_p^1(a,b) \overset{c}{\subset} C[a,b]$.

□ By (14.17) any bounded in $W_p^1(a,b)$ space is uniformly bounded and equipotentially continuous in $C[a,b]$; therefore, it is compact by the Artsel theorem. ■

§ 15. Generalized Solution of the Dirichlet Problem for Elliptic Equations of the Second Order

Classic Dirichlet problem. Integral identity. Generalized problem. Approximate methods: Galerkin methods and their realization in the finite element method.

1. Classic Dirichlet problem. Integral identity

Let $\Omega \subset \mathbf{R}^n$ be an open bounded domain with sufficiently smooth boundary $\partial\Omega$. Take the following boundary value problem in its classic statement, called the *Dirichlet problem*: find a function $u(x) \in C^2(\Omega) \cap C(\overline{\Omega})$ that is a solution to the equation

$$-\sum_{i,j=1}^{n} \frac{\partial}{\partial x_j}\left(a_{ij}(x)\frac{\partial u}{\partial x_j}\right) + a(x)u = f(x) \tag{15.1}$$

under the boundary condition

$$u|_{\partial\Omega} = 0. \tag{15.2}$$

We assume that the functions $a(x), f(x) \in C(\overline{\Omega})$ with $a(x) \geq 0$, $a_{ij}(x) = a_{ji}(x) \in C^1(\overline{\Omega})$, $i, j = \overline{1, n}$, and there exists a constant $\beta > 0$ such that the inequality

$$\sum_{ij=1}^{n} a_{ij}(x)\xi_i\xi_j \geq \beta \sum_{i=1}^{n} \xi_i^2 \tag{15.3}$$

holds for any vector $\xi = (\xi_1, \xi_2, \ldots, \xi_n)$ and $\forall\, x \in \bar{\Omega}$.

Existence and uniqueness of a solution to problem (15.1) and (15.2) are proven in the theory of partial differential equations.

Let $u(x)$ be a solution of problem (15.1) and (15.2) and let $v \in C_0^1(\bar{\Omega})$, then $v|_{\partial\Omega} = 0$. Multiply both parts of Equation (15.1) by the function v and integrate with respect to the domain Ω:

$$\int_\Omega \left(-v \sum_{i,j} \frac{\partial}{\partial x_i} \left(a_{ij} \frac{\partial u}{\partial x_j} \right) + auv \right) d\Omega = \int_\Omega f v d\Omega.$$

Apply the derivatives $\partial/\partial x_i$ to the function v, integrating by parts, to obtain

$$\int_\Omega \left(\sum_{i,j=1}^{n} a_{ij} \frac{\partial u}{\partial x_j} \frac{\partial v}{\partial x_i} + auv \right) d\Omega = \int_\Omega f v d\Omega \tag{15.4}$$

$$\forall\, v \in C_0^1(\overline{\Omega}).$$

Integrals with respect to the boundary $\partial\Omega$ disappear during this transformation due to the condition $v|_{\partial\Omega} = 0$. Since equality (15.4) is valid for $\forall\, v \in C_0^1(\overline{\Omega})$, it is called the *integral identity.*

2. Generalized problem. Existence and uniqueness of its solution

Now let us generalize the statement of the Dirichlet boundary value problem under essentially weaker conditions for the equation coefficients $a_{ij}(x), a(x)$ and the right hand part $f(x)$. Let $u, v \in \overset{\circ}{W}{}_2^1(\Omega)$; take the bilinear form:

$$a(u, v) = \int_\Omega \left(\sum_{i,j=1}^{n} a_{ij} \frac{\partial u}{\partial x_j} \frac{\partial v}{\partial x_i} + auv \right) d\Omega, \tag{15.5}$$

where the coefficients $a_{ij}(x)$, $a(x)$ are bounded piecewise continuous functions, $a_{ij} = a_{ji}$ satisfy the condition (15.3), and $a(x) \geq 0$. It is continuous in $\overset{\circ}{W}{}_2^1(\Omega)$ and $\overset{\circ}{W}{}_2^1(\Omega)$-elliptic, since by (15.3) for $\forall\, u \in \overset{\circ}{W}{}_2^1(\Omega)$

$$a(u, v) \geq \beta \int\limits_\Omega \sum_{i=1}^n \left(\frac{\partial u}{\partial x_i}\right)^2 d\Omega = \beta \|u\|^2_{\overset{\circ}{W}{}_2^1(\Omega)}. \tag{15.6}$$

Let $f \in L_2(\Omega)$, then $f(v) = \int\limits_\Omega f v d\Omega$ is linear bounded on the elements $v \in \overset{\circ}{W}{}_2^1(\Omega)$ functional, since by the Cauchy–Bunyakovskii and the Poincaré–Steklov inequalities (14.10)

$$\left|\int f v\, d\Omega\right|^2 \leq \int\limits_\Omega f^2 a\Omega \int\limits_\Omega v^2 d\Omega \leq C \int\limits_\Omega f^2 d\Omega \|v\|^2_{\overset{\circ}{W}{}_2^1(\Omega)}. \tag{15.7}$$

We call a *generalized solution* of problem (15.1) and (15.2) under the preceding conditions for the functions $a_{ij}(x)$, $a(x)$, $f(x)$ the function $u(x) \in \overset{\circ}{W}{}_2^1(\Omega)$ for which integral identity (15.4) is satisfied $\forall v \in \overset{\circ}{W}{}_2^1(\Omega)$ where the integrals are the Lebesgue integrals.

This definition of the generalized solution of problem (15.1) and (15.2) is equivalent to the definition of a solution to the following problem stated in the form of a variational equation: find a function $u \in \overset{\circ}{W}{}_2^1(\Omega)$ such that

$$a(u, v) = f(v) \quad \text{for} \quad \forall\, v \in \overset{\circ}{W}{}_2^1(\Omega), \tag{15.8}$$

where $a(u, v)$ is assigned with formula (15.5) and $f(v)$ is bounded by virtue of the (15.7) linear functional. By the Vishik–Lax–Milgram theorem a solution to problem (15.8) [or (15.4)] exists and is unique. Obviously, any classic solution of problem (15.1) and (15.2) is a generalized solution too, though, the inverse statement is, generally speaking, incorrect.

Remark. We could take the Dirichlet problem for the more general form of Equation (15.1), according to the chosen way of reasoning, adding the sum $\sum\limits_{i=1}^n b_i(x)(\partial u/\partial x_i)$ to the left part and the sum

$\sum_{i=1}^{n}(\partial/\partial x_i)F_i(x)$ to the right part, where $F_i(x) \in L_2(\Omega)$. If corresponding bilinear form $a(u,v)$ (nonsymmetric already) is bounded and $\overset{\circ}{W}{}_2^1$-elliptic, then the conclusion on the existence and uniqueness of the generalized solution [as a solution to Equation (15.8)] stays valid.

3. Approximate Galerkin and Ritz methods and their realization in the finite element method

So, let generalized problem (15.8) be posed; the symmetry of $a(u,v)$ is not obligatory. Let $H_n \subset \overset{\circ}{W}{}_2^1(\Omega)$ be a subspace in $\overset{\circ}{W}{}_2^1(\Omega)$ of a dimension n. Then the generalized Galerkin method is reduced to the substitution of problem (15.8) with the following one: find an element $u_n \in H_n$ satisfying variational equation

$$a(u_n, v_n) = f(v_n) \quad \text{for} \quad \forall\, v_n \in H_n. \tag{15.9}$$

By Theorem 2 of Section 12, the solution u_n of problem (15.9) exists and is unique; we call it a *discrete solution.*

In the case of symmetric bilinear form, the discrete solution is characterized by the property

$$u_n = \arg \inf_{v_n \subset H_n} J(v_n), \tag{15.10}$$

where the quadrature functional

$$J(v) = a(v,v) - 2f(v). \tag{15.11}$$

The algorithm for the discrete solution as a solution to problem (15.10) is called the *Ritz method.*

Let us examine how problem (15.9) has been solved. Let $\{e_k\}_1^n$ be a basis in space H_n. Then the solution

$$u_n = \sum_{k=1}^{n} \bar{u}_k e_k$$

of problem (15.9) is such that its coefficients $\bar{u}_k$ are a solution of the linear system

$$\sum_{k=1}^{n} a(e_k, e_i)\bar{u}_k = f(e_i), \qquad 1 \le i \le n \tag{15.12}$$

whose matrix is always reversible, since bilinear form $a(u, v)$ is elliptic.

It is important, from the standpoint of calculations, that the choice of the basis $\{e_k\}_1^n$ provides for an as much as possible large number of zeros in the resulting matrix and a condition number of this matrix does not deteriorate catastrophically for $n \to \infty$. Solvability of the matrix can be reached, for example, with the special choice of a basis: so-called *finite functions* are taken as e_k (see functions of this type in Figure 4, Section 6 of Chapter 1); in this case the coefficient $a(e_k, e_i)$ turns to zero each time an intersection measure of supports of basis functions e_k and e_i is equal to zero. The Galerkin method using finite basis functions is called the *finite element method* or, if finite functions are piecewise linear, the *variation-difference method.*

The convergence of u_n to the generalized solution u for $n \to \infty$ and the estimation of an error $u - u_n$ are established in the symmetric case on the basis of the inequality (see Section 11):

$$a((u - u_n), (u - u_n)) = \inf_{v_n \in H_n} a(u - v_n, u - v_n). \tag{15.13}$$

The system of equations (15.12) for large n is usually solved with iteration methods.

Chapter 3

Iteration Methods for the Solution of Operator Equations

§ 1. General Theory of Iteration Methods

Method of successive approximation for linear equations of the second kind, necessary and sufficient conditions for convergence, error estimate, sufficient conditions for convergence. Iteration methods for solution of linear equations of the first kind. Definition of multistep, one-step, cyclic linear iteration methods. General form of one-step linear iteration methods. Convergence rate.

Iteration methods are a rather universal tool both for the investigation of existence and uniqueness of solutions for operator equations, and for actual search for these solutions.

1. Method of successive approximation for linear equations of the second kind, necessary and sufficient conditions for convergence, error estimate, sufficient conditions for convergence

Let an operator $T \in \mathcal{L}(\mathbf{X})$ and let $\mathbf{X}$ be a Banach space. *Equations of the second kind* are the following equations

$$u = Tu + \psi, \tag{1.1}$$

where ψ is assigned and u desired elements of the space $\mathbf{X}$.

One of the most frequently used methods for the solution of Equation (1.1) is the *method of successive approximations* or the *method of simple iterations*; having assigned an arbitrary element $u^0 \in \mathbf{X}$ called an *initial approximation* we construct, starting with it, a sequence $\{u^k\}$ of approximate solutions with the formula

$$u^{k+1} = Tu^k + \psi, \qquad k = 0, 1, \dots . \tag{1.2}$$

If the sequence (1.2) proves to be convergent, then we say that the process of successive approximations for Equation (1.1), started with the element u^0, *converges*. Since T is a linear bounded operator, then the fact of the convergence of the sequence $\{u^k\}$ implies that $u = \lim\limits_{k\to\infty} u^k$ is a solution of Equation (1.1). In order to check it, it is sufficient to pass to a limit for $k \to \infty$ in formula (1.2).

The convergence of the method of successive approximations for Equation (1.1) is associated with the contraction mapping principle and the validity of this principle depends on the convergence with respect to a norm of the series

$$\sum_{k=0}^{\infty} T^k u^0, \qquad \forall\, u^0 \in \mathbf{X} \tag{1.3}$$

whose sum (in the case of convergence) is $(I - T)^{-1}u^0$.

Theorem 1. *The method of successive approximations* (1.2) *for Equation* (1.1) *converges for any initial approximation* $u^0 \in \mathbf{X}$ *and for any fixed element* $\psi \in \mathbf{X}$ *to the unique solution* u *of Equation* (1.1) *iff for any* $u^0 \in \mathbf{X}$ *the series* (1.3) *converges. In case of convergence, we have the following estimates for* $\varepsilon^k = u - u^k$

$$\|\varepsilon^k\| = \|T\varepsilon^{k-1}\| \le \|T\|\,\|\varepsilon^{k-1}\| \tag{1.4}$$

$$\|\varepsilon^k\| = \|T^k\varepsilon^0\| \le \|T^k\|\,\|\varepsilon^0\| \tag{1.5}$$

$$\|\varepsilon^k\| = \|(I-T)^{-1}T^k(u^1 - u^0)\| \le \|(I-T)^{-1}T^k\|\,\|u^1 - u^0\|. \tag{1.6}$$

□ Apply formula (1.2) sequentially to obtain

$$u^k = \psi + T\psi + \ldots + T^{k-1}\psi + T^k u^0. \tag{1.7}$$

Let the series (1.3) converge; that is, $T^k u^0 \to 0$ for $k \to \infty$ and there exists

$$u = \lim_{k\to\infty} u^k = \sum_{k=0}^{\infty} T^k\psi = (I-T)^{-1}\psi. \tag{1.8}$$

Then $Tu = \sum_{k=1}^{\infty} T^k\psi$ and, consequently, $u - Tu = \psi$, $u = (I-T)^{-1}\psi$; that is, u is a solution to Equation (1.1).

Let us show the uniqueness of the solution we have found. Suppose that Equation (1.1) has another solution $x \neq u$. Then $v = x - u \neq 0$ and $v = Tv$; consequently, $v = Tv = \ldots = T^k v = \ldots$. There exists a limit for the sum

$$\sum_{k=0}^{n} T^k v = (n+1)v$$

for $n \to \infty$ which implies that $v = 0$. The contradiction proves the uniqueness.

The inverse statement is evident, since if the method of successive approximations (1.2) converges for $\forall u^0 \in \mathbf{X}$ to the solution determined by formula (1.8), then $T^k u^0 \to 0$ for $k \to \infty$ and by (1.7) and (1.8) the series (1.3) converges for any fixed $\psi \in \mathbf{X}$.

Estimate (1.4) can be obtained by the subtraction from the left and right parts of Equation (1.1) corresponding parts of equality (1.2); as a result, we have

$$\varepsilon^{k+1} = T\varepsilon^k. \tag{1.9}$$

It follows from (1.9) that

$$\varepsilon^k = T^k\varepsilon^0. \tag{1.10}$$

Estimate (1.5) follows from (1.10). Finally, since

$$(I-T)\varepsilon^0 - u - Tu - u^0 + Tu^0 = \psi + Tu^0 - u^0 = u^1 - u^0$$

then $\varepsilon^0 = (I-T)^{-1}(u^1 - u^0)$. Substitute this expression for ε^0 into (1.10) to obtain

$$\varepsilon^k = T^k(I-T)^{-1}(u^1 - u^0) = (I-T)^{-1}T^k(u^1 - u^0). \tag{1.11}$$

Inequality (1.6) follows immediately. ∎

Let us notice that if the condition

$$\|T\| = q < 1 \tag{1.12}$$

holds, then the estimates (1.5) and (1.6) cand be replaced with

$$\|\varepsilon^k\| \leq q^k\|\varepsilon^0\|$$

$$\|\varepsilon^k\| \leq q^k(1-q)^{-1}\|u^1 - u^0\|.$$

So, condition (1.12) is a sufficient condition for the convergence of the method of successive approximations; another sufficient condition [as we had established before in Chapter 2, the Neumann series (1.3) converges absolutely if it is valid] is the condition

$$\mu(T) = q < 1. \tag{1.13}$$

Let $\{\varphi_i\}$ be a countable system of eigenelements of an operator T that forms a basis in $\mathbf{X}$ and let $\{\lambda_i\}$ be corresponding eigenvalues:

$$T\varphi_i = \lambda_i\varphi_i, \qquad i = 1, 2, \ldots .$$

Then $|\lambda_i| \leq q < 1$ for all $i \geq 1$ if the condition (1.13) holds. If we take

$$u^k = \sum_{i=1}^{\infty} u_i^k \varphi_i, \qquad \psi = \sum_{i=1}^{\infty} \psi_i\varphi_i, \qquad \varepsilon^k = \sum_{i=1}^{\infty} \varepsilon_i^k \varphi_i, \tag{1.14}$$

where u_i^k, ψ_i, ε_i^k are numerical coefficients in expansions (1.14), then it is easily seen that substituting (1.14) into formulas (1.2), (1.7), and (1.9)–(1.11), and equating corresponding numerical coefficients at each φ_i we get

$$u_i^{k+1} = \lambda_i u_i^k + \psi_i = (1 + \lambda_i + \ldots + \lambda_i^k)\psi_i + \lambda_i^{k+1} u_i^0 \tag{1.15}$$

$$\varepsilon_i^{k+1} = \lambda_i \varepsilon_i^k = \lambda_i^{k+1}\varepsilon_i^0 = \lambda_i^{k+1}(1 - \lambda_i)^{-1}(u_i^1 - u_i^0). \tag{1.16}$$

These formulas demonstrate how approximate solution and errors of iteration method change component by component.

2. Iteration methods for the solution of linear equations of the first kind. Definition of multistep, one-step, cyclic linear iteration methods. General form of one-step linear iteration methods. Convergence rate

Any iteration method for the solution of the *equation of the first kind*

$$Au = f, \tag{1.17}$$

where $u, f \in \mathbf{X}$, $A, A^{-1} \in \mathcal{L}(\mathbf{X})$, implies the determination of a sequence of approximations $u^k \in \mathbf{X}$ that is supposed to be convergent to $A^{-1}f$ for $k \to \infty$. During an iteration of an order r, a formula for the approximation of u^{k+1} is regarded as dependent explicitly on $A, f, u^k, \ldots, u^{k-r+1}$; these are so-called *r-step* iteration methods. Usually they take $r = 1, 2$. We call a method of the form

$$u^{k+1} = F_k(A, f, u^k) \tag{1.18}$$

the *one-step iteration method.* Iteration (1.18) is termed *stationary* if a form of operator function $F_k(\cdot, \cdot, \cdot,)$ does not depend on k. If a form of the function $F_k(\cdot, \cdot, \cdot,)$ changes in a cyclic way with respect to k with a period N, then the iterations are referred to as *cyclic with the period N*. Obviously, such an iteration is equivalent to some stationary iteration as related to the approximations $u^0, u^N, u^{2N}, \ldots$ with a function $F_N(A, f, F_{N-1}(A, f, F_{N-2}(A, f, \ldots) \ldots))$. If F_k is a linear function of u^k, then the iteration is termed *linear.* Linear iterations are the simplest for the investigations.

A function in the form $T_k u^k + \psi^k$, where $T_k = T_k(A, f)$ is some linear operator and ψ^k some element from $\mathbf{X}$, is the most general linear with respect to u^k function F_k. If we demand that such an operation F_k leave the exact solution of problem (1.1) as a stationary point, that is,

$$A^{-1} f = T_k A^{-1} f + \psi^k \tag{1.19}$$

be valid, then we obtain that ψ^k, f and T_k must be connected with the relationship $\psi^k = H_k f$, where $H_k = (I - T_k)A^{-1}$.

So, it is possible to conclude what is the general principle of the construction of iteration methods using only information on the preceding approximation and leaving the exact solution u as a stationary point: as a final result, a sequence of elements u^k is constructed starting with some element u^0 with the formulas

$$u^{k+1} = T_k u^k + \psi^k \tag{1.20}$$

or

$$u^{k+1} = u^k - H_k(Au^k - f), \tag{1.21}$$

where $T_k = I - H_k A$, $\psi^k = H_k f$, and H_k is some sequence of operators characterizing a *type of iteration method.* The operator T_k is called the *transition operator* at the kth step of the iteration method.

An *error* $\varepsilon^k = u - u^k$ where u is an exact solution of Equation (1.17) satisfies the relationship

$$\varepsilon^{k+1} = T_k \varepsilon^k \tag{1.22}$$

that is obtained if one subtracts corresponding parts of (1.20) from the left and right parts of Equation (1.19)

$$u = T_k u + \psi^k,$$

which holds for any k. It follows from (1.22) that

$$\varepsilon^{k+1} = M_k \varepsilon^0, \quad \text{where} \quad M_k = T_k T_{k-1} \cdot \ldots \cdot T_0 \tag{1.23}$$

which means that the convergence of iterations (1.20) for assigned initial error ε^0 depends on the behaviour of $M_k \varepsilon^0$ for $k \to \infty$; iterations (1.20) converge for prescribed initial error ε^0 iff $M_k \varepsilon^0 \to 0$ for $k \to \infty$.

Having substituted $\varepsilon^{k+1} = u - u^{k+1}$ and $\varepsilon^0 = u - u^0$ into (1.23) we obtain a formula for u^{k+1} through initial approximation u^0

$$u^{k+1} = M_k u^0 + (I - M_k) A^{-1} f. \tag{1.24}$$

A value of ε^0 is, as a rule, unknown; it is easier therefore to investigate the behaviour of a *discrepancy*

$$r^k = f - Au^k \tag{1.25}$$

since it can be calculated with the unknown exact solution $u = A^{-1} f$. As a matter of fact, $r^k = Au - Au^k = A\varepsilon^k$, therefore we have by (1.22): $r^{k+1} = AT_k A^{-1} r^k$ or

$$r^{k+1} = AM_k A^{-1} r^0. \tag{1.26}$$

Let an operator $T_k = T$ be independent of k. Since $\|T^k\|^{1/k} = \mu(T) b(k)$ where $b(k) \to 1$ for $\mu(T) > 0$ and $k \to \infty$, then by (1.23)

$$\|\varepsilon^k\| \leq \|T^k\| \, \|\varepsilon^0\| = \mu^k(T) b^k(k) \|\varepsilon^0\|. \tag{1.27}$$

Let us write, instead of (1.27), asymptotic for $k \to \infty$ equality $\|\varepsilon^k\| / \|\varepsilon^0\| \sim (b(k) \cdot \mu(T))^k$; taking a logarithm of it, we obtain that in order to decrease a norm of the error $\|\varepsilon^k\|$ as compared to $\|\varepsilon^0\|$ by $1/\varepsilon$ times, it is sufficient for $0 < \mu(T) < 1$ and sufficiently small $\varepsilon > 0$ to perform

$$k \approx \ln \varepsilon / \ln \mu(T) \tag{1.28}$$

iterations. This formula is asymptotic one; it can be refined for n-dimensional operators with the help of $b(k)$. It happens, that for $k \to \infty$, $\ln b(k) = ((d-1)/k) \ln k$, where d, $1 \leq d \leq n$ is the maximal order for a Jordan box of the matrix T. Formula (1.28) is incorrect for nilpotent operators.

The value $-\ln \mu(T)$ is called the *asymptotic degree of convergence* of iterations or simply the *degree of convergence*. The closer is $\mu(T)$ to zero, the higher is the degree of convergence.

§ 2. On the Existence of Convergent Iteration Methods and Their Optimization

Methods for the transformation of an equation to the form convenient for iterations; symmetrizable case. On existence of iteration methods for solution of linear equations of the first kind. Methods accelerating convergence. Optimization problem. Loss functional. Stationary case.

1. Methods for the transformation of an equation to the form convenient for iterations; symmetrizable case

In Section 1 we found when the iteration method (1.20) converges. It is not clear yet if there exist operators H_k that provide for the convergence of iterations (1.20) for the solution of Equation (1.17) and what the conditions are when these operators H_k exist. Before we start to investigate this question, let us consider possible types of transformations that reduce the equation of the first kind (1.17) into the equation of the second kind (1.1). We perform such transformations usually with the following combinations:

(a) adding the same element to both parts of (1.17) or the partially transformed equation;

(b) applying the same operator to both parts of (1.17) or the partially transformed equation (multiplying from the left by an operator or scalar);

(c) substituting the unknown u with the formula $u = Qy$.

The first transformation reduces an equation into an equivalent one. For the second, if the linear bounded operator H is applied, any solution of the initial equation is a solution for the new equation. If there exists an inverse operator, the inverse statement is valid. It is necessary for the substitution of the unknown $u = Qy$ that the operator Q be continuously invertible.

Let us show some simple types of transformations. Multiply the equation $Au - f = 0$ from the left by an operator D, then add to both sides an element $-Cu$ where C is some operator. Then Equation (1.17) transforms into the equation

$$Cu = Cu - D(Au - f). \tag{2.1}$$

Consequently, the iteration method can be written as

$$Cu^{k+1} = Cu^k - D(Au^k - f). \tag{2.2}$$

This method corresponds, if the operator C has an inverse one, to the method (1.21) with $H = C^{-1}D$.

Similarly, in order to solve the equation

$$u = Bu + f \tag{2.3}$$

with an iteration method, multiply it by an operator K

$$Ku = KBu + Kf \tag{2.4}$$

and add to both parts a member $-Lu$. Assuming that there exists the operator $(K - L)^{-1}$ we arrive at the equation

$$u = (K - L)^{-1}(KB - L)u + (K - L)^{-1}Kf. \tag{2.5}$$

This equation, in the case of successful choice of operators K, L, can be taken as an equation for the iteration method (1.20). It is possible, in turn, to apply to Equations (2.1) and (2.5), if appropriate, the preceding transformations. The operators C, D, K, L can depend in these equations on k.

Let us describe one more type of transformation useful for both the construction of satisfactorily convergent iteration methods and for the investigation of their convergence degree.

Let H be a Hilbert space with an inner product $(\cdot,\cdot)$ and let H_M be a Hilbert space with an inner product $[\cdot,\cdot]_M = (M\cdot,\cdot)$ with positive definite symmetric operator M (see Section 10 of Chapter 2). Let $\mathbf{X}$, $\mathbf{Y}$ be Banach spaces and let the operator $L \in \mathcal{L}(\mathbf{X},\mathbf{Y})$ have an inverse operator L^{-1}. Assign an equation to be transformed in the form

$$Lv = g, \tag{2.6}$$

where $g \in \mathbf{Y}$ and $v \in \mathbf{X}$ is desired solution.

Let the operator L be (C,D)-symmetrizable in H_M (see item 4, Section 10 of Chapter 2); then transform, with the help of substitutions

$$u = Dv, \qquad f = Cg, \qquad A = CLD^{-1}, \tag{2.7}$$

Equation (2.6) into Equation (1.17) with selfadjoint positive definite in H_M operator A. Denote by M_A, m_A, respectively, upper and lower bounds for the operator A.

Apply to transformed Equation (1.17) iteration methods (see, e.g., Section 5). The method transforming Equation (2.6) into Equation (1.17) is sometimes called the *preconditioning*, meaning that we obtain from (2.6) an equation with a selfadjoint positive definite operator that possesses the least condition measure ($K(A) = M_A/m_A < K(L)$).

2. On the existence of iteration methods for the solution of linear equations of the first kind

Again let $A \in \mathcal{L}(\mathbf{X})$; the following theorem is valid.

Theorem 1. *Equation* (1.17) *has a unique solution u for each $f \in \mathbf{X}$ iff there exists a continuously invertible operator H possessing a property such that the series*

$$\sum_{k=0}^{\infty} T^k g, \tag{2.8}$$

where $T = I - HA$ converges for any $g \in \mathbf{X}$. The solution u in this case is given by the formula

$$u = \sum_{k=0}^{\infty} T^k H f.$$

□ Let there exist an operator H possessing the preceding properties. Take $g = Hf$. By virtue of the completeness of $\mathbf{X}$ and our assumption, the series (2.8) converges to some element $u \in \mathbf{X}$ and $u = \sum_{k=0}^{\infty} T^k g$. Then $Tu = \sum_{k=1}^{\infty} T^k g = u - g$ or $u - HAu = u - g$; that is,

$$HAu = Hf. \tag{2.9}$$

Multiplying (2.9) from the left by H^{-1} we see for ourselves that u is a solution to Equation (1.17). The uniqueness has been proved by contradiction. Let $\bar{u}$ be another solution of (1.17) and let $v = u - \bar{u}$. Then $Av = 0$, $HAv = 0$; that is, $(I - T)v = 0$ where $T = I - HA$. It means that $v = Tv = T^k v$ for $\forall\, k$, hence, as in a similar theorem from Section 1 of this chapter, $v = 0$. The contradiction proves the uniqueness.

Notice in order to prove the inverse statement of the theorem, if Equation (1.17) has a unique solution $u \in \mathbf{X}$ for each $f \in \mathbf{X}$, then the operator A is continuously invertible. Then it is possible to choose $H = \eta A^{-1}$ where η is a number such that $|1 - \eta| < 1$. More than that, H^{-1} exists and is equal to $\eta^{-1}A$ and $T = I - HA = (1 - \eta)I$. Since $\|T\| = |1 - \eta| < 1$, then by Corollary 2 from Section 4 of Chapter 2 the series $\sum_{k=0}^{\infty} T^k g = \sum_{k=0}^{\infty} (1 - \eta)^k g$ converges to some element for any g. Denote by $\bar{u}$ a sum of this series for $g = Hf$. Since $\sum_{k=0}^{\infty} (1 - \eta)^k g = \eta^{-1} g$, then $\bar{u} = \eta^{-1} g = \eta^{-1} H f = A^{-1} f = u$. ∎

So, it is necessary and sufficient for the existence of operators H providing convergent iteration methods for the solution of Equation (1.17), that this equation would have a unique solution for any $f \in \mathbf{X}$. It is seen from the proof of this theorem that operator A^{-1} plays an exceptional role in the construction of iteration processes, namely, if $H = A^{-1}$, then after one iteration with formula (1.20) we obtain at once the exact solution no matter what initial approximation u^0 is taken. This operator is not the only one providing a convergent process. The theorem on perturbations also determines some set of operators where operators H can be chosen, namely:

$$\{H : \|H - A^{-1}\| < \|A\|^{-1}\} \tag{2.10}$$

since the series (2.7) converges for such choice of H. Convergence conditions for the iterations are determined also by Theorem 1 from Section 1 and by conditions (1.12) and (1.13).

Remark. An efficient algorithm for the calculation of partial sums of the Neumann series in (2.8) is presented in Section 10 of this chapter.

3. Methods accelerating the convergence. Optimization problem. Loss functional. Stationary case

What is the optimal way to select the operator H? We have seen that if $H = A^{-1}$, then the exact solution is obtained with only one iteration. It is obvious that, generally speaking, it is not an optimal way of solution, since an enormous number of arithmetic operations must be spent in finding operator A^{-1}. If an operator of simpler structure is taken as the operator H, then it is possible that there exists an iteration process requiring a smaller number of operations during one iteration. If this process converges, then using it to solve problem (1.17) with accuracy ε [spending $k(\varepsilon)$ iterations where $k(\varepsilon)$

is determined with formula (1.28)] we obtain an approximate solution with accuracy ε for a smaller number of operations as compared to the case $H = A^{-1}$.

Let us try to select a criterion for the comparison of efficiency of iteration processes for the solution of the same problem. To this end, we formalize and simplify to some extent an actual situation, neglecting some facts whose importance can actually be essential in some cases.

It is evident that the convergence degree of the iteration method (1.20) characterized for $T_k = T$ by a value $\mu(T)$ and the number of operations for one iteration cannot be considered as two independent characteristics while estimating the quality of the method. Let us assume that most of the operations in each iteration of iteration method (1.20) are spent calculating an element like Tv with assigned element v. Let $C(T)$, a *cost of operation* T, that is, a value characterizing the working time for the calculation of the element Tv, be proportional to the total number of reduced arithmetic and logical operations necessary for the calculation of the element Tv through any element v.

Take now convergent iteration method (1.2). Let $v^k = u^{2k}$, then

$$v^{k+1} = TTv^k + (I+T)\psi \tag{2.11}$$

and this method converges with the degree $-\ln\mu(T^2)$, that is, roughly twice as fast as method (1.2). Suppose computer memory volume does not permit the calculation of the operator T^2v in any way except the method of sequential double multiplication of the element v by the operator T. Let us calculate the asymptotic number of operations for $k \to \infty$ spent in the methods (1.2) and (2.11) for the solution of the problem with accuracy ε: these are the values $C(T)\ln\varepsilon/\ln\mu(T)$ and $C(T^2)\ln\varepsilon/\ln\mu(T^2) \sim C(T)\ln\varepsilon/\ln\mu(T)$, respectively, since $\mu(T^2) = \mu^2(T)$ and $C(T^2) = 2C(T)$. So, although method (2.10) has a higher degree of convergence, we get no profit with respect to time using it as compared to method (1.2).

The example with methods (1.2) and (2.11) was intentionally taken as simple in order to demonstrate the essence of the problem: it seems that creating a new method we should be careful of the value

$$W(T) = -C(T)/\ln\mu(T) \tag{2.12}$$

which must be sufficiently small. Let us call this value a *loss functional.* Generally speaking, any modifications of "old" iteration methods are reasonable when they (under the same or other conditions) decrease the loss functional.

The optimization problem for a given family of stationary iteration methods is much simpler in the case when the operators H (and consequently T) depend on scalar parameters so that the value $C(T)$ stays constant. In this case the method with a higher degree of convergence is better. It is equivalent to the statement that a method with the least spectral radius of transition operator is better.

Take one of the simple problems of this kind. Let $\mathbf{X} = H$ be a Hilbert space. Let us regard Equation (1.17) to be solved as an equation obtained as a result of equivalent transformations in a form convenient for the application of an iteration method of type

$$u^{k+1} = u^k - \alpha(Au^k - f), \tag{2.13}$$

where α is a scalar; so, $H = \alpha I$ in (2.13).

As for the operator A, we assume here and in the next sections (if there is no special remark) that A is a selfadjoint positive definite operator whose spectrum ($Sp(A)$) belongs to a segment $[m, M]$, $0 < m < M$ and the points m and M belong to $Sp(A)$ and that the operator A has a countable complete in H orthonormed system of eigenelements φ_i, $i = 1, 2, \dots$

$$A\varphi_i = \lambda_i\varphi, \qquad i = 1, 2, \dots \qquad (\lambda_i \in [m, M]). \tag{2.14}$$

In method (2.13), $T = I - \alpha A$; consequently, elements φ_i are eigenelements of the operator T and

$$T\varphi_i = \nu_i(\alpha)\varphi_i, \qquad i = 1, 2, \dots\ , \tag{2.15}$$

where $\nu_i(\alpha) = 1 - \alpha\lambda_i$. We see, with regard to formula (9.9) of Chapter 2 for the spectral radius of an operator through a spectrum of the operator, that the optimization problem for the iteration method (2.13) is equivalent to the problem of the search for

$$\inf_{\alpha} \sup_{\lambda \in Sp(A)} |1 - \alpha\lambda|. \tag{2.16}$$

Since linear with respect to λ function $1-\alpha\lambda$ takes extremal values at the ends of a segment $[m, M]$ and the points m, M are extreme points of a spectrum, the problem (2.16) is equivalent to the problem

$$\inf_{\alpha} \max\{|1 - \alpha m|, |1 - \alpha M|\}. \tag{2.17}$$

We notice now that the value α, as a solution to problem (2.17), must be such that the values in $\{,\}$ would be equal, since otherwise, changing α, their maximal could be decreased so that it would still stay maximal. So, maximal α must satisfy the equation

$$|1-\alpha m| = |1-\alpha M|. \tag{2.18}$$

The value $\alpha = 0$ will not do; consequently, $1-\alpha m = -(1-\alpha M)$; that is,

$$\alpha = \alpha_{\text{opt}} = \frac{2}{M+m}. \tag{2.19}$$

For this value of α

$$\mu(T) = 1 - \alpha_{\text{opt}} m = \frac{M-m}{M+m} < 1 \tag{2.20}$$

or

$$\mu(T) = \frac{1-\frac{m}{M}}{1+\frac{m}{M}}. \tag{2.21}$$

So, we have obtained an optimal iteration method of type (2.13) where the optimal value of a parameter is given by formula (2.19) and the convergence rate is estimated with the value $\mu(T)$ expressed by formula (2.21).

Since the operators A and A^{-1} are selfadjoint, then $\|A\| = M$, $\|A^{-1}\| = m^{-1}$; so, the condition measure $K(A) = M/m$. We see by formula (2.21) that $\mu(T) \to 1$ for $m/M \to 0$, that is, the degree of convergence of the method (2.13) slows down ($-\ln \mu(T) \to 0$) with the growth of the condition number $K(A)$.

Notice finally, that we regarded the value α in method (2.13) as not changing from iteration to iteration, and to determine α_{opt} we used just the information about the two extreme points of the spectrum. The location of the spectrum inside $[m, M]$ does not affect the solution of optimization problem (2.16).

§ 3. The Chebyshev One-Step (Binomial) Iteration Methods

Formulation of a cyclic method, optimality. Error estimation. Formulas for parameters. Notion of stability and algorithms for ordering of parameters. Examples for $N = 2^p, 3^p$. Infinitely continuable stable optimal methods.

The Chebyshev polynomials are broadly used in the optimization of iteration methods for the solution of linear inhomogeneous equations and partial eigenvalue problems. We call iteration methods, where properties or parameters of the Chebyshev polynomials are used, the *Chebyshev methods.* As for operator A of problem (1.17), we assume that it satisfies the statements formulated in the previous section.

1. Formulation of a cyclic method, optimality. Error estimation. Formulas for parameters

In order to find a solution u to Equation (1.17), take the following iteration method with varying parameters. Let some integer $N > 0$ be given and let α_i be scalars, $i = \overline{1, N}$; perform N iterations with formulas

$$u^{k+1} = u^k - \alpha_{k+1}(Au^k - f), \qquad k = 0, 1, \ldots, N-1, \tag{3.1}$$

where u^0 is prescribed. We need to select parameters α_i, $i = \overline{1, N}$ so that after N iterations we obtain the best (in a certain sense) convergence of iterations. If it is necessary to continue the iteration process (3.1), then we can believe that parameters in such a method are repeated with period N; that is, $\alpha_{k+N} = \alpha_k$. So, we arrive at a cyclic method with period N.

The method (3.1) has the transition operator $T_k = (I - \alpha_{k+1}A)$. After N iterations (3.1), using relationship (1.24), we get

$$u^{k+N} = P_N(A)u^k + (I - P_N(A))A^{-1}f, \tag{3.2}$$

where $P_N(t)$ is a polynomial of power N that has the form

$$P_N(t) = \prod_{i=1}^{N}(1 - \alpha_i t), \qquad P_N(0) = 1 \tag{3.3}$$

and the following recursion relationships are valid for the errors $\varepsilon^k = u - u^k$ [see (1.23)]

$$\varepsilon^{k+N} = P_N(A)\varepsilon^k. \tag{3.4}$$

Set $k = 0$ in (3.4) to obtain an estimate for an error after N iterations through the norm of initial error:

$$\|\varepsilon^N\| = \|P_N(A)\varepsilon^0\| \le \|P_N(A)\| \, \|\varepsilon^0\|. \tag{3.5}$$

Taking into account that the operators A and $P_N(A)$ are selfadjoint and that $\mathrm{Sp}(A) \subset [m, M]$ we obtain

$$\|P_N(A)\| = \mu(P_N(A)) = \sup_{\lambda \in \mathrm{Sp}(A)} |P_N(\lambda)| \le \max_{\lambda \in [m,M]} |P_N(\lambda)|. \tag{3.6}$$

Select the coefficients α_i, $i = \overline{1, N}$ in method (3.1) so that the polynomial $P_N(t)$ of type (3.3) would be the least deviating from zero on the segment $[m, M]$. Then we obtain the *optimal method* for the whole class of problems of type (1.17) with selfadjoint operators A whose spectrum belongs to the segment $[m, M]$. The search for the desired polynomial is equivalent to the problem of finding the values

$$E_N = \inf_{\alpha_i} \max_{t \in [m,M]} |P_N(t)| \tag{3.7}$$

and the latter is reduced easily to problem I (c) discussed in Section 10 of Chapter 1. In fact, we see after the substitution of variables

$$x = \frac{M + m - 2t}{M - m} \tag{3.8}$$

that the points M, m, 0 on the axis t pass, respectively, into the points -1, 1 and

$$\theta = \frac{M + m}{M - m} > 1 \tag{3.9}$$

on the axis x. In Section 10 of Chapter 1 we solved the problem of the construction of a polynomial of Nth power the least deviating from zero on $[-1, 1]$ and taking the value 1 in a point $x = \theta > 1$. The polynomial $T_N(x)/T_N(\theta)$ was such a polynomial, where $T_N(x)$ is a Chebyshev polynomial of the first kind.

Consequently, coming back to old variables with formula (3.8) we see that the polynomial

$$P_N(t) = \frac{1}{T_N(\theta)} \cdot T_N\left(\frac{M + m - 2t}{M - m}\right) \tag{3.10}$$

is the desired polynomial of Nth power the least deviating from zero on the segment $[m, M]$ that is equal to the unit for $t = 0$.

For this polynomial $E_N = 1/T_N(\theta)$; that is, formula (3.5) takes the form

$$\|\varepsilon^N\| \leq \frac{1}{T_N(\theta)} \|\varepsilon^0\|. \tag{3.11}$$

Let us calculate $T_N(\theta)$ with formula (10.9) from Chapter 1:

$$T_N(\theta) = \frac{1}{2}((\theta + \sqrt{\theta^2 - 1})^N + (\theta + \sqrt{\theta^2 - 1})^{-N}).$$

It is easily seen that $\theta + \sqrt{\theta^2 - 1} = \sigma^{-1}$ where

$$\sigma = \frac{M^{1/2} - m^{1/2}}{M^{1/2} + m^{1/2}} = \frac{1 - \sqrt{\frac{m}{M}}}{1 + \sqrt{\frac{m}{M}}} < 1; \tag{3.12}$$

consequently,

$$\|\varepsilon^N\| \le \frac{2\sigma^N}{1+\sigma^{2N}}\|\varepsilon^0\|. \tag{3.13}$$

Let us compare this formula to the similar formula for the iteration method with constant α_{opt} presented in Section 2. We have for method (2.13) after N iterations:

$$\|\varepsilon^N\| \le \left(\frac{1-\frac{m}{M}}{1+\frac{m}{M}}\right)^N \|\varepsilon^0\|. \tag{3.14}$$

Consequently, for $N > 1$

$$\frac{2\sigma^N}{1+\sigma^{2N}} < \left(\frac{1-\frac{m}{M}}{1+\frac{m}{M}}\right)^N. \tag{3.15}$$

It is easy to show that for small $\xi = m/M > 0$ and large N the left-hand part of inequality (3.15) is much larger than the right-hand part.

Let us find now formulas for parameters of method (3.1). By (3.3) α_i^{-1} are the roots of the polynomial $P_N(t)$, but the roots of the polynomial $P_N(t)$ are easily found with formula (3.10) through the roots β_j of the polynomial $T_N(x)$

$$\left(\beta_j = \cos\psi_j\pi, \quad \psi_j = \frac{2j-1}{2N}, \quad j = \overline{1,N}\right).$$

Take for each root β_j a parameter α_k according to the following rule. Let

$$\kappa_N = (j_1, j_2, \ldots, j_N) \tag{3.16}$$

be an integer permutation of an order N $(1 \le j_k \le N,\ j_i \ne j_k)$. Take in the method (3.1)

$$\alpha_k = 2(M + m - (M-m)\cos\psi_{j_k}\pi)^{-1}, \tag{3.17}$$

where

$$\psi_j = \frac{2j-1}{2N} \qquad j = \overline{1,N}. \tag{3.18}$$

So, the *parameters for optimal method* (3.1) *are chosen with formulas* (3.17) and (3.18) *and used in the order determined by the permutation* κ_N (3.16). Then inequality (3.13) is valid after N iterations (3.1) for the error ε^N.

2. The notion on stability and algorithms for the ordering of parameters. Examples for $N = 2^p,\ 3^p$

The *stability* of the method (3.1) and (3.16)–(3.18) is an important problem. Its stability as related to roundoff errors depends essentially on the usage order of parameters (permutations k_N), since for small values of $\xi = m/M$ and of numbers j_k the transition operators $I - \alpha_k A$ will have an enormous norm. It may lead, in real calculations with computers, to two undesirable consequences: catastrophic growth of $\|u^k\|$ for some $1 \le k \le N$ which may result in overflow or in the loss of significant digits; and similar growth of errors at intermediate iterations. Let

$$R_i^N(t) = \prod_{j=1}^{i}(1 - \alpha_j t), \qquad Q_i^N = \prod_{j=i+1}^{N}(1 - \alpha_j t) \tag{3.19}$$

$$r_i^N = \max_{t\in[m,M]} |R_i^N(t)|, \qquad q_i^N = \max_{t\in[m,M]} |Q_i^N(t)|. \tag{3.20}$$

Take a simple case: we have made an error ξ_i just on the ith iteration in the process (3.1); that is, instead of (3.1), performed the calculation with formula

$$u^i = u^{i-1} - \alpha_i(Au^{i-1} - f) + \xi_i$$

for $k = i$. Express u^i through u^0 to obtain [see (1.24)]

$$u^i = R_i^N(A)u^0 + (I - R_i^N(A))A^{-1}f + \xi_i. \tag{3.21}$$

Continue the iteration process to obtain instead of formula (3.2), the expression

$$u^N = P_N(A)u^0 + (I - P_N(A))A^{-1}f + Q_i^N(A)\xi_i. \tag{3.22}$$

Formulas (3.21) and (3.22) show that there will be no essential loss of significant digits due to the growth of $\|u^i\|$ if $r_i^N \le C_1$, $i = \overline{1, N}$ since r_i^N gives an upper estimate of a norm of transition operator in method (3.1) for the first i iterations and calculations will be stable as related to errors that appear at the ith iteration if $q_i^N < C_2$, $i = \overline{1, N}$ since q_i^N gives an upper estimate of a norm of transition operator in method (3.1) from the ith to Nth iteration; here C_1 and C_2 are constants depending only on $\xi = m/M$. It is necessary besides, in order to obtain a good general estimate of an error accumulating for

N iterations in method (3.1) due to roundoff errors, that the following values also satisfy similar inequalities

$$\sum_{i=1}^{N} q_i^N, \quad \sum_{i=1}^{N} q_i^N r_i^N. \tag{3.23}$$

It is obvious that the values r_i^N, q_i^N for small ξ depend essentially on a type of permutation κ_N and the operators $I - \alpha_i A$ with a large norm must in stable methods be distributed in a sufficiently uniform way among operators that decrease a norm of error.

Let us formulate a problem. Find for polynomials $P_N(t)$ of the type (3.10) the sequences $\{N_p\}_1^\infty$ $(N_p < N_{p+1})$ and permutations κ_{N_p} such that for $N = N_p$ and any $1 \leq i \leq N$ the values (3.20) and (3.23) in representation (3.19) would be bounded by constants depending just on $\xi = m/M$.

We say that a set N of different numbers $\Psi = \{\psi_i\}_1^N$, enumerated in increasing order, is ordered by permutation $\kappa_N = (j_1, j_2, \ldots, j_N)$ if from Ψ a new sequence $\Omega = \{\omega_k\}_1^N$ is obtained according to the rule

$$\omega_k = \psi_{jk}, \qquad k = 1, 2, \ldots, N. \tag{3.24}$$

If the set Ψ consists of m not intersecting subsets $\Psi = \bigcup_{l=1}^{m} \Psi_l$, then we say that the elements of Ψ are partially ordered by the permutation $\bar{\kappa}_m = (j_1, j_2, \ldots, j_m)$ if the new enumeration is introduced on elements of Ψ such that just elements of the set Ψ_{j1} are assigned the first numbers, then elements of the set Ψ_{j2}, and so forth.

Let

$$S_k(x, \beta) = \frac{T_k(x) - \cos\beta}{T_k(\theta) - \cos\beta}, \tag{3.25}$$

where x and $\theta > 1$ are defined by formulas (3.8) and (3.9) and let $0 \leq \beta \leq \pi$. The polynomial of kth power $S_k(x, \beta)$ has on $[-1, 1]$ equal maximums and equal minimums and for $\forall\, \theta > 1$

$$\max_{x \in [-1,1]} |S_k(x, \beta)| \leq 1 \quad \text{for} \quad \frac{\pi}{2} \leq \beta \leq \pi. \tag{3.26}$$

Let us obtain the useful factorizations $S_{2k}(x, \beta)$ and $S_{3k}(x, \beta)$.

By formulas (10.3) and (10.11) from Chapter 1

$$S_{2k}(x, \beta) = S_k\left(x, \pi - \frac{\beta}{2}\right) \cdot S_k\left(x, \frac{\beta}{2}\right). \tag{3.27}$$

By (3.26) the first multiplier in (3.27) does not exceed the unit with respect to a module.

Taking into account that roots of the equation

$$T_3(x) = \cos\beta$$

are found explicitly (simple trigonometric equation $\cos 3\varphi = \cos\beta$ must be solved) and are expressed as

$$y_1 = \cos\tfrac{1}{3}(2\pi+\beta) < 0, \quad y_2 = \cos\tfrac{1}{3}(2\pi-\beta), \quad y_3 = \cos\tfrac{\beta}{3} > 0, \tag{3.28}$$

where

$$y_2 > 0 \text{ for } \frac{\pi}{2} < \beta \leq \pi \quad \text{and} \quad y_2 \leq 0 \text{ for } 0 \leq \beta \leq \frac{\pi}{2} \tag{3.29}$$

we have

$$S_{3k}(x,\beta) = S_k\left(x, \frac{1}{3}(2\pi+\beta)\right) S_k\left(x, \frac{1}{3}(2\pi-\beta)\right) S_k\left(x, \frac{\beta}{3}\right); \tag{3.30}$$

moreover,

$$\max_{x\in[-1,1]} |S_k(x, \frac{1}{3}(2\pi+\beta))| \leq 1. \tag{3.31}$$

It is easily shown that

$$\max_{x\in[-1,1]} |S_k(x, \frac{1}{3}(2\pi+\beta)) \cdot S_k(x, \frac{1}{3}(2\pi-\beta))| \leq 1 \tag{3.32}$$

and

$$\max_{x\in[-1,1]} |S_k(x, \frac{1}{3}(2\pi-\beta))| \leq 1 \quad \text{for} \quad 0 \leq \beta \leq \frac{\pi}{2}. \tag{3.33}$$

Formula (3.10) takes in new notation the form

$$P_N(t) = S_N\left(x, \frac{\pi}{2}\right); \tag{3.34}$$

the set $\Psi = \Psi_N$ consists of numbers (3.18) that belong to $[0,1]$ and $x_j = \cos\psi_j\pi$ are the roots of $T_N(x)$. We order the set Ψ_N and say about this set that its elements form the roots x_j.

We present here, skipping a complete proof, two simple recursive algorithms developed by this author and S. Finogenov for the construction of κ_N in the stable optimal method (3.1). Let

$N = N_p = 2^p$ $(p > 0)$, $\kappa_1 = (1)$ and let $\kappa_{2^{p-1}} = (j_1, j_2, \ldots, j_{2^{p-1}})$ be a permutation of the desired algorithm for $N = 2^{p-1}$. Then define the permutation κ_{2^p} with the formula

$$\kappa_{2^p} = (2^p + 1 - j_1, j_1, 2^p + 1 - j_2, j_2, \ldots); \tag{3.35}$$

for example: $\kappa_{16} = (11, 6, 14, 3, 10, 7, 15, 2, 12, 5, 13, 4, 9, 8, 16, 1)$.

The following factorization of $P_{2^p}(t)$ corresponds to permutation (3.35) according to (3.34) and (3.27)

$$P_{2^p}(t) = S_{2^p}\left(x, \frac{\pi}{2}\right) = S_{2^{p-1}}\left(x, \frac{3}{4}\pi\right) \cdot S_{2^{p-1}}\left(x, \frac{\pi}{4}\right). \tag{3.36}$$

We see that the first 2^{p-1} members of permutation (3.35) correspond to the roots of the first multiplier in (3.36) and the last 2^{p-1} members to the roots of the second multiplier. It means that the set Ψ is partially ordered, being divided into two parts Ψ_1 and Ψ_2 (during the enumeration, elements of Ψ_1 precede elements of Ψ_2). Factorizing in turn each multiplier in (3.36) with formula (3.27) into two, we partition each of the sets Ψ_1, Ψ_2 into subsets; the first one consists of ψ_j that have corresponding roots of the first multiplier, and the second one consists of ψ_j that have corresponding roots of the second multiplier in (3.27). We take here that ψ_j from the first subset always precede ψ_j from the second one. Continuing this partition process we have finally all ψ_j ordered with permutation (3.35).

Let $N = 2^p$ and let the number $1 \leq i < 2^p$ have representation in a binary system (it is unique)

$$i = \sum_{k=1}^{l} 2^{m_k}, \tag{3.37}$$

where $0 \leq m_1 < m_2 < \cdots < m_l < p$; then [see (3.19) and (3.27)]

$$R_i^N(t) = \prod_{k=1}^{l} S_{2^{m_k}}\left(x, \pi - \frac{\beta_{m_k}}{2}\right), \tag{3.38}$$

where $0 \leq \beta_{m_k} \leq \pi$. In fact, if a multiplier of the type $S_{2^{m_k}}(x, \beta_{m_k}/2)$ were present in (3.38), then the multiplier (as the preceding one) $S_{2^{m_k}}(x, \pi - \beta_{m_k}/2)$ would be present too. That is, according to (3.27), a multiplier with index 2^{m_k+1} would be in (3.38) instead of multiplier with index 2^{m_k} which contradicts representation (3.38). So, we obtain with regard to inequality (3.26) that

$$r_i^N = \max_{t \in [m, M]} |R_i^N(t)| \leq 1, \qquad i = 1, 2, \ldots, N. \tag{3.39}$$

The second algorithm is taken for the sequence $N = N_r = l \cdot 3^r$ where l is fixed and $r = 0, 1, \ldots$ We have by formulas (10.11) and (10.21) of Chapter 1:

$$T_{l\cdot 3^r}(x) = 2T_{l\cdot 3^{r-1}}(x)\left(T_{3^{r-1}}(T_{2l}(x)) - \cos\frac{\pi}{3}\right); \tag{3.40}$$

consequently,

$$P_N(t) = S_N\left(x, \frac{\pi}{2}\right) = S_{l\cdot 3^{r-1}}\left(x, \frac{\pi}{2}\right) \cdot S_{3^{r-1}}\left(T_{2l}(x), \frac{\pi}{3}\right). \tag{3.41}$$

Apply, if possible, a similar transformation to the first multiplier in (3.41) and then to the first multipliers of the transformed product to obtain

$$P_N(t) = S_1\left(T_l(x), \tfrac{\pi}{2}\right) S_1\left(T_{2l}(x), \tfrac{\pi}{3}\right) S_3\left(T_{2l}(x), \tfrac{\pi}{3}\right)$$
$$\ldots S_{3^k}\left(T_{2l}(x), \tfrac{\pi}{3}\right) \ldots S_{3^{r-1}}\left(T_{2l}(x), \tfrac{\pi}{3}\right). \tag{3.42}$$

Formula (3.41) shows that all roots of the polynomial $S_{l\cdot 3^{r-1}}(x, \pi/2)$ are the roots of the polynomial $S_{l\cdot 3^r}(x, \pi/2)$. Assume that this group of roots precedes the rest of the roots $S_N(x, \pi/2)$; the values ψ_j corresponding to this group will not change and get new numbers passing from $l \cdot 3^{r-1}$ to $l \cdot 3^r$. More exactly, we assume that if $\kappa_{l\cdot 3^{r-1}} = (\bar{j}_1, \bar{j}_2, \ldots, \bar{j}_{l\cdot 3^{r-1}})$ is a permutation of an order $l \cdot 3^{r-1}$ already constructed, then the first third of indices in the permutation

$$\kappa_{l\cdot 3^r} = (j_1, j_2, \ldots, j_{l\cdot 3^r}) \tag{3.43}$$

is determined with the formulas: $j_k = 3\bar{j}_k - 1$, $k = 1, 2, \ldots, l\cdot 3^{r-1}$. We determine two thirds of the indices left in (3.43) ordering the values ψ_j corresponding to the roots of the second multiplier in (3.41). Let us establish the following rule for partial ordering inside some groups of ψ_j that appear in the transformation process. Let it be that at some stage of the factorization of $S_{3^{r-1}}(T_{2l}(x), \pi/3)$ we have obtained a multiplier of the type $S_{3k}(T_{2l}(x), \beta)$ for some β determined by the preceding transformations and let it have some corresponding set ψ_j; denote it conditionally by $\Psi_k(\beta)$. According to formula (3.30)

$$S_{3k}(T_{2l}(x), \beta) = S_k(T_{2l}(x), \frac{1}{3}(2\pi + \beta))$$
$$\cdot S_k(T_{2l}(x), \frac{1}{3}(2\pi - \beta)) \cdot S_k\left(T_{2l}(x), \frac{\beta}{3}\right). \tag{3.44}$$

Let $\Psi_k(\beta) = \bigcup_{i=1}^{3} \Psi_k^i(\beta)$ where the set Ψ_k^i corresponds to roots of the ith parenthesis in (3.44). *A rule is as follows*: if $0 \leq \beta \leq \pi/2$, order the set $\Psi_k(\beta)$ partially with permutation (2,1,3), and if $\pi/2 < \beta \leq \pi$, then with permutation (1,2,3). According to this rule we conclude with regard to inequalities (3.31)–(3.33), that the first third of renumerated ψ_j from $\Psi_k(\beta)$ will always correspond to a multiplier in (3.44) that is not larger than the unit by a module, and the first two thirds of renumerated ψ_j will correspond to the first two multipliers in (3.44), and a module of their product is not larger than the unit.

So, an algorithm for the determination of the last indices in permutation (3.43) is defined for $r \geq 2$ by a chain of the following transformations with the application at each stage of the preceding rule for partial ordering. We obtain at the first stage

$$S_{3^r-1}\left(T_{2l}(x), \tfrac{\pi}{3}\right) = S_{3^r-2}\left(T_{2l}(x), \tfrac{7}{9}\pi\right)$$
$$\cdot S_{3^r-2}\left(T_{2l}(x), \tfrac{5}{9}\pi\right) \cdot S_{3^r-2}\left(T_{2l}(x), \tfrac{\pi}{9}\right) \qquad (3.45)$$

and apply permutation (2,1,3). At the second stage for $r \geq 3$ factorize each of the multipliers in (3.45) with formula (3.44) and order the values ψ_j according to our rule. Continuing this process we finally reach multipliers of $S_1(T_l(x), \pi/2$ and $S_1(T_{2l}(x), \beta)$; type and order the values ψ_j for them with permutations κ_l, κ_{2l}, respectively. We determine thus all the elements of permutation (3.43). Notice that the obtained succession order of groups of ψ_j corresponds to the succession order of multipliers in formula (3.42).

Let us assume further that $l = 2^p$; then permutations of this type have been defined before. Take, as an example, $p = 0$, $r = 2$; then $\kappa_9 = (5, 8, 2, 7, 3, 6, 4, 9, 1)$.

If permutations (3.35) are taken as κ_{2^p}, then it is possible to show that for $N_r = 2^p \cdot 3^r$, inequalities (3.39) hold for r_i^N.

The estimations for values q_i^N are performed in both algorithms in a more complicated way; we do not present them here. Notice, though, that in the algorithms we constructed the largest norm of transition operator belongs to the last step of method (3.1) for small values of $\xi = m/M$. This norm has an upper bound

$$q_{N-1}^N = \frac{2\cos^2 \frac{\pi}{4N_r}}{\theta - 1 + 2\sin^2 \frac{\pi}{4N_r}}$$

that grows for $r \to \infty$ as

$$\left(\xi + \frac{\pi^2}{16N_r^2}\right)^{-1}$$

and tends to M/m.

Remark 1. If parameters α_{k+1} are interpreted in method (3.1) as steps t_0^{k+1} in the methods of minimization of functionals (see Section 7), then the suggested algorithm for the choice of steps for $N = 2^p \cdot 3^r$ imitates successfully descent tactics in the so-called "ravine" methods for the solution of ill-conditioned linear minimization problems: first they perform a series of small steps of various length in order to descend to the bottom of a ravine, and then a big conclusive step is made along the ravine's bottom.

3. Infinitely continuable stable optimal methods

The iteration method (3.1) and (3.16)–(3.18) becomes optimal only after N steps of iterations and, generally speaking, does not provide for optimal approximation at intermediate iterations. A new class of iteration processes allows us to continue method (3.1) after N iterations so that it would be stable for some sequence $\{N_r\}$ ($N \le N_r < N_{r+1}, N_r \to \infty$ for $r \to \infty$) too and become optimal again at the N_lth iteration step.

Let us present an algorithm for the construction of such a method for the partial case when $N_r = 2^p \cdot 3^r$. To this end, construct an infinite sequence of numbers $\omega_k \in [0,1]$. Instead of formula (3.42) formally write the following infinite product

$$S_\infty(x) = S_1\left(T_{2^p}(x), \frac{\pi}{2}\right) \prod_{i=0}^{\infty} S_{3^i}\left(T_{2^{p+1}}(x), \frac{\pi}{3}\right). \tag{3.46}$$

Taking r multipliers under the product sign and applying the algorithm presented in item 2, we obtain the ordered set $\{\omega_k\}_1^{N_r}$. Adding to the product one more multiplier, raise this set with $2N_r$ members to obtain the ordered sequence $\{\omega_k\}^{N_{r+1}}$. Continuing to raise the number of members of the sequence in the similar way, we obtain an infinite sequence of numbers $\{\omega_k\}_1^\infty$. It is possible to show that it is uniformly distributed on $[0,1]$ and such that the first numbers $\{\cos\omega_k\pi\}$ for $k \le 3^r 2^p$ are the roots of the polynomial $T_{2^p \cdot 3^r}(x)$. Let us call the constructed sequence a $\{\omega_k\}_1^\infty$ *T-sequence.* For example, for $N_r = 2\cdot 3^r$ a T-sequence of constructed type has the form

$$\frac{3}{4}, \frac{1}{4}, \frac{7}{12}, \frac{5}{12}, \frac{11}{12}, \frac{1}{12},$$
$$\frac{23}{36}, \frac{13}{36}, \frac{31}{36}, \frac{5}{36}, \frac{25}{36}, \frac{11}{36}, \frac{29}{36}, \frac{7}{36}, \frac{19}{36}, \frac{17}{36}, \frac{35}{36}, \frac{1}{36}, \dots$$

If we now take

$$\alpha_k = 2(M + m - (M - m)\cos\omega_k\pi)^{-1} \tag{3.47}$$

in method (3.1), then we obtain a stable method that becomes optimal for all $k = 2^p\cdot 3^r$, $r = 0, 1, \dots$; for this method

$$r_i^{N_r} \le 1, \qquad i = 1, 2, \dots, \qquad r = 0, 1, \dots . \tag{3.48}$$

§ 4. The Chebyshev Two-Step (Trinomial) Iteration Method

Formulation of a two-step iteration method and associated system of polynomials. Optimization, formulas for optimal parameters.

1. Formulation of two-step (trinomial) iteration method and associated system of polynomials

Let us apply the following iteration method to the solution of Equation (1.17) where operator A satisfies the conditions formulated in Section 2

$$u^{k+1} = u^k - \alpha_{k+1}(Au^k - f) - \beta_{k+1}(u^k - u^{k-1}), \qquad k = 0, 1, \dots ; \tag{4.1}$$

here $\beta_1 = 0$, u^0 is given, and α_{k+1}, β_{k+1} are numerical parameters to be found. For $k \ge 1$, this method demands storing two preceding approximations in the computer memory.

Let $\varepsilon^\alpha = u - u^\alpha$. Having expressed from this relationship the relationships u^{k-1}, u^k, u^{k+1} for $\alpha = k$, $k-1$, $k+1$ and substituted them into (4.1) we obtain the relationships that are satisfied with the error

$$\varepsilon^1 = \varepsilon^0 - \alpha_1 A\varepsilon^0 \tag{4.2}$$

and

$$\varepsilon^{k+1} = (I - \beta_{k+1} I - \alpha_{k+1} A)\varepsilon^k + \beta_{k+1}\,\varepsilon^{k-1} \tag{4.3}$$

for $k \geq 1$. One can see from these formulas that ε^{k+1} may be expressed through ε^0 and that this connection has the form

$$\varepsilon^{k+1} = P_{k+1}(A)\varepsilon^0, \tag{4.4}$$

where $P_{k+1}(A)$ is a polynomial from operator A of power $k+1$. Determine a form of this operator. By equality (4.4), formally taking there $k+1=0$, obtain that $P_0(t)=1$. It follows from (4.2) that $P_1(t) = 1 - \alpha_1 t$ and substituting (4.4) into (4.3) we get the relationships

$$P_{k+1}(A)\varepsilon^0 = (I - \beta_{k+1} I - \alpha_{k+1} A)P_k(A)\varepsilon^0 + \beta_{k+1} P_{k-1}(A)\varepsilon^0$$

which means that the polynomials $P_{k+1}(t)$, $P_k(t)$, and $P_{k-1}(t)$ are interconnected with the dependence

$$P_{k+1}(t) = (1-\beta_{k+1}-\alpha_{k+1}t)P_k(t)+\beta_{k+1}P_{k-1}(t),\ k=1,2,\dots. \tag{4.5}$$

All polynomials $P_k(t)$ satisfy the condition

$$P_k(0) = 1 \tag{4.6}$$

since $P_0(0) = P_1(0) = 1$ and by (4.5)

$$P_{k+1}(0) = (1-\beta_{k+1})P_k(0) + \beta_{k+1}P_{k-1}(0), \qquad k = 1,2,\dots.$$

So, we have established that each iteration method (4.1) may be related with a system of polynomials $P_k(t)$ satisfying condition (4.6) and recursive relationships

$$P_{k+1}(t) = (1 - \beta_{k+1} - \alpha_{k+1}t)P_k(t) + \beta_{k+1}P_{k-1}(t), \tag{4.7}$$

where $\beta_1 = 0$ and $P_0(t) = 1$.

2. Optimization. Formulas for optimal parameters

Let us determine parameters for an optimal method. Demand, to this end, that for any $k \geq 1$ the polynomials $P_k(t)$ satisfy condition (4.6) and, being determined by relationships (4.7), would at the same time be the polynomials the least deviating from zero on the segment $[m, M]$. Then by (3.10)

$$P_k(t) = \frac{T_k(x)}{T_k(\theta)}, \tag{4.8}$$

where x and θ are obtained with formulas (3.8), and (3.9). Determine from (4.8) $T_i(x)$ for $i = k-1,\ k,\ k+1$ and substitute these values and quantity x from formula (3.8) into the relationships

$$T_1(x) = x, \qquad T_{k+1}(x) = 2xT_k(x) - T_{k-1}(x), \qquad k = 1, 2, \dots \tag{4.9}$$

that are satisfied with Chebyshev polynomials, to obtain that

$$T_1(\theta)P_1(t) = \frac{M + m - 2t}{M - m}$$

$$T_{k+1}(\theta)P_{k+1}(t) = 2\left(\frac{M + m - 2t}{M - m}\right) T_k(\theta)P_k(t) - T_{k-1}(\theta)P_{k-1}(t).$$

By comparison of these formulas with formulas (4.7) we obtain that

$$\beta_1 = 0, \qquad \alpha_1 = \frac{2}{M + m} \tag{4.10}$$

and

$$\alpha_{k+1} = \frac{4}{M - m}\frac{T_k(\theta)}{T_{k+1}(\theta)}, \qquad \beta_{k+1} = -\frac{T_{k-1}(\theta)}{T_{k+1}(\theta)}. \tag{4.11}$$

With these values of parameters method (4.1) becomes optimal for each $k \geq 1$ since by formulas (4.4) and (4.8)

$$\|\varepsilon^k\| \leq \frac{1}{T_k(\theta)} \|\varepsilon^0\| \tag{4.12}$$

or

$$\|\varepsilon^k\| \leq \frac{2\sigma^k}{1 + \sigma^{2k}} \|\varepsilon^0\|, \tag{4.13}$$

where σ is determined by formula (3.12).

It is easier to calculate the coefficients $\alpha_{k+1}, \beta_{k+1}$ for $k \geq 1$ with the following recursive formulas. Let $\delta_{k+1} = T_k(\theta)/T_{k+1}(\theta)$. Then

$$\alpha_{k+1} = 4\delta_{k+1}/(M - m), \qquad \beta_{k+1} = -\delta_k\, \delta_{k+1} \tag{4.14}$$

but according to (4.9) for $k \geq 1$ and $x = \theta$, $\delta_{k+1}^{-1} = 2\theta - \delta_k$; that is,

$$\delta_{k+1} = (2\theta - \delta_k)^{-1}, \qquad \delta_1 = \theta^{-1}, \qquad k = 1, 2, \dots . \tag{4.15}$$

So, computing the values $\alpha_{k+1}, \beta_{k+1}$ with formulas (4.10), (4.14), and (4.15) we arrive at method (4.1) that for any k decreases a norm of error in the best way.

Let us estimate an important characteristic of methods (3.1) and (4.1) such as the average degree of convergence, which is understood naturally for the kth iteration as the value

$$V(k) = \frac{1}{k} \ln T_k(\theta). \tag{4.16}$$

Lemma 1. *The value $V(k)$ is a function monotonically increasing with k; $\lim\limits_{k\to\infty} V(k) = -\ln\sigma$.*

□ According to (3.12)

$$T_k(\theta) = \sigma^{-k} \frac{(1 + \sigma^{2k})}{2}$$

then $V(k) = -\ln\sigma + W(k)$; $W(k) = (1/k)\ln(1 + \sigma^{2k}/2) < 0$ here. Since $0 < \sigma < 1$ then, obviously, $\lim\limits_{k\to\infty} W(k) = 0$. In order to prove the monotonic growth of $W(k)$ perform the substitution of variables:

$$\sigma^{2k} = 1 - t \quad \text{or} \quad 2k\ln\sigma = \ln(1-t) \qquad (0 < t < 1).$$

When k changes from 0 to ∞, t changes from 0 to 1. Then

$$W(k) = \overline{W}(t) = 2\ln\sigma \ln\left(1 - \frac{t}{2}\right) / \ln(1-t).$$

Show, assuming t as the variable value, that $\overline{W}(t)$ is a monotonically increasing function of t in the interval (0.1). Having taken a derivative of $\overline{W}(t)$

$$\overline{W}'(t) = \frac{-(1 - t/2)\ln(1 - t/2) + 1/2(1-t)\ln(1-t)}{(1-t)(1-t/2)\ln^2(1-t)}(-2\ln\sigma)$$

we see that $\ln\sigma < 0$ there and the denominator is positive. The lemma will be proven if we show that the numerator is positive for $0 < t < 1$ also. Notice to this end, that it turns to zero for $t = 0$ and possesses a positive derivative equal to $(1/2)\ln(1 - t/2)(1 - t)$. ■

§ 5. The Chebyshev Iteration Methods for Equations with Symmetrized Operators

Formulas for iteration methods, convergence estimations, choice of symmetrization operators.

Let H, H_M be Hilbert spaces with inner products $(\cdot,\cdot)$ and $[\cdot,\cdot]_M = (M\cdot,\cdot)$, respectively, where $M > 0$; $\mathbf{X}$, $\mathbf{Y}$ be Banach spaces; an operator $L \in \mathcal{L}(\mathbf{X},\mathbf{Y})$ have inverse one L^{-1}, $g \in \mathbf{Y}$, and let the equation

$$Lv = g \tag{5.1}$$

be given for the determination of an element $v \in \mathbf{X}$.

Let the operator L be (C,D)-symmetrizable in H_M (see item 4 in Section 10 of Chapter 2); then convert Equation (5.1) with substitution (2.7) into Equation (1.17)

$$Au = f \tag{5.2}$$

with selfadjoint positive definite in H_M operator A whose spectrum belongs to the segment $[m_A, M_A]$, $0 < m_A \le M_A$.

In order to find a solution to Equation (5.2) we apply the Chebyshev binomial (3.1) and trinomial (4.1) iteration methods where obtained approximations are designated by u^k and errors by $\varepsilon^k = u - u^k$. If the substitution

$$v^k = D^{-1}u^k \tag{5.3}$$

is used in formulas of methods (3.1) and (4.1), then being applied to Equation (5.1) they will look, respectively, like

$$v^{k+1} = v^k - \alpha_{k+1}D^{-1}C(Lv^k - g), \qquad k = 0,1,\dots,N-1; \tag{5.4}$$

$$v^{k+1} = v^k - \alpha_{k+1}D^{-1}C(Lv^k - g) - \beta_{k+1}(v^k - v^{k-1}), \tag{5.5}$$

$$\beta_1 = 0, \qquad k = 0,1,\dots .$$

Recall that parameters in methods (5.4) and (5.5) are defined with formulas (3.17), and (4.10) and (4.11) for $m = m_A$, $M = M_A$ and that relationships (3.4), (4.2), and (4.3) are valid for the errors ε^k as well as estimates (3.13) and (4.13) where norms are defined in the space H_M: $\|\varepsilon^k\| = [\varepsilon^k]_M = (M\varepsilon^k,\varepsilon^k)^{1/2}$. If $\eta^k = v - v^k$, then we have, respectively, for methods (5.4) and (5.5) $\eta^k = D^{-1}\varepsilon^k$ and

$$\eta^{k+1} = \eta^k - \alpha_{k+1}D^{-1}CL\eta^k, \tag{5.6}$$

$$\eta^{k+1} = \eta^k - \alpha_{k+1} D^{-1} C L \eta^k - \beta_{k+1}(\eta^k - \eta^{k-1}). \tag{5.7}$$

By formula (3.13) we obtain after N iterations for method (5.4)

$$[D\eta^N]_M \leq \frac{2\sigma^N}{1+\sigma^{2N}}[D\eta^0]_M \tag{5.8}$$

and by formula (4.13) for method (5.5) and any k

$$[D\eta^k]_M \leq \frac{2\sigma^k}{1+\sigma^{2k}}[D\eta^0]_M, \tag{5.9}$$

where σ is determined by formula (3.12) for $m = m_A$, $M = M_A$.

There is no exact prescription for the choice of operators C and D except for two ideal but impractical cases: $C = L^{-1}$, $D = I$ or $C = I$, $D = L$ and known Gauss transformation: $C = L^*$, $D = I$ that can worsen the condition number for an operator of transformed Equation (5.2).

Let $\mathbf{X} = \mathbf{Y} = H$ and $D(M) = H$. Denote by a star the operators adjoint in H and by $B_i \in \mathcal{L}(H)$ auxiliary operators possessing inverse ones. Set off the following technique and its possibility for the construction of symmetrizing operators.

1. First, let $L^* = L > 0$. For this case, the transformation of Equation (5.1) into Equation (5.2) means improvement for the conditionality of the transformed equation. Take

$$C = B_1^{*-1}, \qquad D = M^{-1}B_1 \tag{5.10}$$

then

$$A = B_1^{*-1} L B_1^{-1} M \tag{5.11}$$

and formulas of methods (5.4) and (5.5) after obvious transformations will look, respectively, like

$$B_1^* M^{-1} B_1 (v^{k+1} - v^k) = -\alpha_{k+1}(Lv^k - g) \tag{5.12}$$

$$B_1^* M^{-1} B_1 (v^{k+1} - v^k + \beta_{k+1}(v^k - v^{k-1})) = -\alpha_{k+1}(Lv^k - g). \tag{5.13}$$

Moreover, according to formula (8.14) of Chapter 2 for $\forall\, v \in H_M$

$$m_A(Mv, v) \leq [Av, v]_M = (LB_1^{-1}Mv, B_1^{-1}Mv) \leq M_A(Mv, v). \tag{5.14}$$

If one takes

$$w = B_1^{-1} M v \tag{5.15}$$

then by (5.14) we obtain for $\forall w \in H_M$ inequalities that must be satisfied by operators B_1, M:

$$m_A[B_1 w, B_1 w]_{M^{-1}} \leq (Lw, w) \leq M_A[B_1 w, B_1 w]_{M^{-1}} \tag{5.16}$$

with the ratio m_A/M_A smaller than the similar ratio for initial Equation (5.1).

2. Now let $L^* \neq L$ or $L^* = L$ but let L not be an operator of fixed sign. Take in this case

$$C = B_1^{*-1} L^* B_2^{*-1} B_2^{-1}, \quad D = M^{-1} B_1 \tag{5.17}$$

then

$$A = B_1^{*-1} L^* B_2^{*-1} B_2^{-1} L B_1^{-1} M \tag{5.18}$$

and formulas of methods (5.4) and (5.5) after transformations will, respectively, take the form

$$B_1^* M^{-1} B_1 (v^{k+1} - v^k) = -\alpha_{k+1} L^* (B_2 B_2^*)^{-1} (Lv^k - g) \tag{5.19}$$

$$B_1^* M^{-1} B_1 (v^{k+1} - v^k + \beta_{k+1} (v^k - v^{k-1}))$$

$$= -\alpha_{k+1} L^* (B_2 B_2^*)^{-1} (Lv^k - g), \tag{5.20}$$

where by the assumption for all $v \in H_M$

$$m_A (Mv, v) \leq [Av, v]_M$$

$$= (B_2^{-1} L B_1^{-1} M v, B_2^{-1} L B_1^{-1} M v) \leq M_A (Mv, v). \tag{5.21}$$

If substitution (5.15) is performed in (5.21), then we obtain for $\forall w \in H_M$ inequalities that must be satisfied with operators B_1, B_2, M:

$$m_A [B_1 w, B_1 w]_{M^{-1}} \leq (B_2^{-1} L w, B_2^{-1} L w)$$

$$\leq M_A [B_1 w, B_1 w]_{M^{-1}}. \tag{5.22}$$

So, we have obtained the iteration methods (5.12), (5.13), (5.19), and (5.20) whose average degree of convergence is estimated by Lemma 1, Section 4 of this chapter with the value $-\ln \sigma$. Efficiency of these methods should be estimated with the loss functional (2.12); $\ln \mu = \ln \sigma$ in its denominator and a cost of one iteration is in its nominator that consists mostly of the values $C(L)$,

$C((B_1^* M^{-1} B_1)^{-1})$ for methods (5.12) and (5.13), and additional values $C(L^*)$, $C((B_2 B_2^*)^{-1})$ for methods (5.19) and (5.20). Consequently, the efficiency of the preceding methods will depend on our skill while solving with respect to any prescribed element w the equation

$$B_1^* M^{-1} B_1 u = w \tag{5.23}$$

for the methods (5.12), (5.13), (5.19), and (5.20) and additionally the equation

$$B_2 B_2^* u = w \tag{5.24}$$

for the methods (5.19) and (5.20).

Equations (5.23) and (5.24) are equations with selfadjoint in H operators. Simplified models of elliptic boundary value problems are used as such equations for the solution of stationary problems in mathematical physics.

§ 6. Block Chebyshev Method

Formulas for the block method, optimization, determination of optimal parameters, estimations of convergence.

1. Formulas for the block method

Iteration methods for so-called *composite problems* have been developed intensively lately; they are called the methods of composition, domain decomposition, bordering, and so forth. Their essence is as follows.

Let the main problem be rather complex but able to be decomposed into parts called subproblems for which there exist effective algorithms of solution. The algorithms we are interested in are iteration algorithms for the solution of the main problem using, essentially, the algorithms for the solution of subproblems and connections between their solutions. For the sake of simplicity, we take the case of two subproblems.

Let it be possible to divide the unknowns in Equation (1.17) that form an element u in two groups: $u = (u_1, u_2)$. The right-hand part of (1.17) will be divided in a similar way: $f = (f_1, f_2)$ (here u and f are two-dimensional vector-columns) and the operator A turns into a matrix of the second order whose elements are operators. As a result operator Equation (1.17) can be written in matrix form:

$$\begin{pmatrix} D_1 & -C_1 \\ -C_2 & D_2 \end{pmatrix} \begin{pmatrix} u_1 \\ u_2 \end{pmatrix} = \begin{pmatrix} f_2 \\ f_2 \end{pmatrix}. \tag{6.1}$$

Assume that D_i are symmetric positive definite operators and that subproblems of the type $D_i v = \psi_i$ are easily solved for them. The operators C_i connect the unknowns u_1 and u_2; we regard them as bounded ones. We do not introduce new spaces whose elements are used for the construction of operators D_i, C_i: a reader will be overloaded otherwise with unnecessary constructions; he or she will understand this from the following text. Let us take the following class of iteration methods: starting with prescribed elements u_1^k, u_2^k, we find the next approximation u_1^{k+1}, u_2^{k+1} with formulas:

$$\begin{aligned} D_1 u_1^{k+1/2} &= C_1 u_2^k + f_1, \\ u_1^{k+1} &= u_1^k + \alpha_{k+1}(u_1^{k+1/2} - u_1^k), \\ D_2 u_2^{k+1/2} &= C_2 u_1^{k+1} + f_2, \\ u_2^{k+1} &= u_2^k + \beta_{k+1}(u_2^{k+1/2} - u_2^k), \end{aligned} \tag{6.2}$$

$$k = 0, 1, \dots ,$$

where u_1^0 and u_2^0 are given. Take just one the of versions of method (6.2), namely: assign $\alpha_{k+1} = 1$. Then $u_1^{k+1} = u_1^{k+1/2}$ and method (6.2) takes the form

$$\begin{aligned} D_1 u_1^{k+1} &= C_1 u_2^k + f_1, \\ D_2 u_2^{k+1/2} &= C_2 u_1^{k+1} + f_2, \\ u_2^{k+1} &= u_2^k + \beta_{k+1}(u_2^{k+1/2} - u_2^k). \end{aligned} \tag{6.3}$$

If $\varepsilon_i^\alpha = u_i - u_i^\alpha$, $\alpha = k, k + (1/2)$, $i = 1, 2$, are errors, then it is easily seen that they satisfy the equations

$$\begin{aligned} D_1 \varepsilon_1^{k+1} &= C_1 \varepsilon_2^k, \\ D_2 \varepsilon_2^{k+1/2} &= C_2 \varepsilon_1^{k+1}, \\ \varepsilon_2^{k+1} &= \varepsilon_2^k + \beta_{k+1}(\varepsilon_2^{k+1/2} - \varepsilon_2^k). \end{aligned} \tag{6.4}$$

Use Equations (6.4) to express ε_2^{k+1} through ε_2^k. We have from the first two Equations (6.4)

$$\varepsilon_2^{k+1/2} = D_2^{-1} C_2 D_1^{-1} C_1 \varepsilon_2^k.$$

Then, if we designate

$$H = D_2^{-1} C_2 D_1^{-1} C_1 \tag{6.5}$$

from the last of Equations (6.4) we have

$$\varepsilon_2^{k+1} = (I + \beta_{k+1}(H - I))\varepsilon_2^k. \tag{6.6}$$

Having performed N steps in method (6.3), we obtain from (6.6) that

$$\varepsilon_2^N = Q_N(H)\varepsilon_2^0, \tag{6.7}$$

where the polynomial Q_N has the form

$$Q_N(t) = \prod_{i=1}^{N}(1 + \beta_i(t - 1)). \tag{6.8}$$

Mmoreover,

$$Q_N(1) = 1. \tag{6.9}$$

Let the operator H have a complete system of eigenelements $\{\varphi_i\}_1^\infty$ in corresponding space that forms the basis

$$H\varphi_i = \lambda_i\varphi_i, \tag{6.10}$$

where $\lambda_i \in [m_0, M_0]$ and $1 \notin [m_0, M_0]$. Then, decomposing ε_2^N and ε_2^0 into the series $\varepsilon_2^N = \sum \varepsilon_i^N \varphi_i$, $\varepsilon_2^0 = \sum \varepsilon_i^0 \varphi_i$, we obtain by (6.7) that

$$\varepsilon_i^N = Q_N(\lambda_i)\varepsilon_i^0 \tag{6.11}$$

2. Optimization, determination of optimal parameters, estimation of convergence

Let us make use of the new optimality criterion. Formula (6.11) determines some linear operator B that maps an element $\bar{\varepsilon}_2^0 = \{\varepsilon_1^0, \dots, \varepsilon_n^0, \dots\} \in m$ onto element $\bar{\varepsilon}_2^N = \{\varepsilon_1^N, \dots, \varepsilon_n^N, \dots\} \in m$. The norm of this element equals $\sup_{\lambda \in Sp(H)} |Q_N(\lambda)|$. Let us select coefficients $\{\beta_i\}_1^N$ of the polynomial $Q_N(\lambda)$ so that they would be solutions of the problem

$$E_N = \inf_{\beta_i} \max_{t \in [m_0, M_0]} |Q_N(t)|. \tag{6.12}$$

The polynomial $Q_N(t)$, for which (6.12) is realized, can be determined easily:

$$Q_N(t) = \frac{1}{|T_N(\theta)|} T_N\left(\frac{M_0 + m_0 - 2t}{M_0 - m_0}\right), \tag{6.13}$$

where

$$\theta = \frac{M_0 + m_0 - 2}{M_0 - m_0}. \tag{6.14}$$

But then

$$\beta_k^{-1} = 1 + \frac{1}{2}\left(M_0 + m_0 - (M_0 - m_0)\cos\omega_{j_k}\bar{u}\right), \tag{6.15}$$

where $\omega_{j_k} = (2j_k - 1)/(2N)$ and κ_N form the permutation $(j_1, j_2, \ldots, j_n)$. More than that,

$$E_N = |T_N(\theta)|^{-1}. \tag{6.16}$$

Consequently, the convergence of a method in the space m can be estimated with inequalities

$$|\varepsilon_i^N| \leq E_N |\varepsilon_i^0|. \tag{6.17}$$

Do not row with just one oar

§ 7. The Descent Methods

Derivative of quadrature functional with respect to direction, gradient, antigradient, methods of gradient descent, gradient method of steepest descent.

We expound in the next two sections on nonlinear iteration methods. We begin with iteration methods for the minimization of quadrature functional $G_M(x)$ (10.1) of Chapter 2 that provide minimizing sequences.

1. Derivative, gradient, antigradient of quadrature functional

Using the notations and assumptions of Section 10 of Chapter 2, take the spaces H and H_M and the quadrature functional $G_M(x)$ (10.1) writing it in the form (10.4). Let $x = w + t\eta$, where $w, \eta \in H_M$,

$H_M \hookrightarrow H$, and let D be the selfadjoint positive definite in H operator; it generates with the inner product $[x,y]_D = (Dx,y)$ for $\forall\, x,y \in H$ the new metric $[x]_D = [x,x]_D^{1/2}$ in H that is called the D-metric. Let us call the expression

$$\left.\frac{\partial G_M(w)}{\partial \eta}\right|_D = \frac{1}{[\eta]_D}\frac{\partial}{\partial t}G_M(w+t\eta)\Big|_{t=0} \tag{7.1}$$

a *derivative* in a point w of the functional $G_M(x)$ with respect to direction η in the D-metric. Having written the functional $G_M(x)$ in the form (10.4) of Chapter 2, we obtain

$$G_M(x) = G_M(w) + 2t[w-u,\eta]_M + t^2[\eta,\eta]_M; \tag{7.2}$$

that is

$$\left.\frac{\partial G_M(w)}{\partial \eta}\right|_D = \frac{2[w-u,\eta]_M}{[\eta]_D} = 2[D^{-1}M(w-u), \frac{\eta}{[\eta]_D}]_D. \tag{7.3}$$

An element η for which the derivative (7.3) takes the largest value is called a *gradient* of the functional $G_M(x)$ in the point w and D-metric; we designate it as $\text{grad}_D G_M(w)$ and call the element $\eta_0 = -\text{grad}_D G_M(w)$ an *antigradient.* It follows from formula (7.3) that with accuracy up to a multiplier, for $w \neq u$

$$\eta_0 = -\text{grad}_D G_M(w) = D^{-1}M(u-w). \tag{7.4}$$

Notice that for $D = M$ and $w \in H_M$

$$w - \text{grad}_M G_M(w) = u. \tag{7.5}$$

Determine now a value $t = t_0$ for which the functional $G_M(w+t\eta_0)$ attains a minimum, descending along a gradient, that is, direction of the most intensive decrease. Differentiate (7.2) with respect to t and equate this derivative to zero for $\eta = \eta_0$ to obtain

$$t_0 = \frac{[\eta_0]_D^2}{[\eta_0]_M^2}. \tag{7.6}$$

Let us elucidate how the functional will change for $t = \omega t_0$ during this process. Notice that $[w-u,\eta_0]_M = [D^{-1}M(w-u),\eta_0]_D$; assign

$$\bar{w} = w + \omega t_0 \eta_0 \tag{7.7}$$

to obtain

$$G_M(\bar{w}) = G_M(w) - \omega \frac{[\eta_0]_D^2}{[\eta_0]_M^2}(2[w-u,\eta_0]_M - \omega[\eta_0]_D^2)$$

$$= G_M(w) - \omega(2-\omega)\frac{[\eta_0]_D^4}{[\eta_0]_M^2}. \quad (7.8)$$

The error functional $\Omega_M(x)$ [see (10.5) in Chapter 2] differs from $G_M(x)$ just with a constant term; therefore,

$$\Omega_M(\bar{w}) = [u-\bar{w}]_M^2 = [u-w]_M^2 - \omega(2-\omega)\frac{[\eta_0]_D^4}{[\eta_0]_M^2}. \quad (7.9)$$

Let $\omega = 1$; by (7.7)

$$D^{-1}M(\bar{w}-u) = D^{-1}M(w-u) + t_0 D^{-1}M\eta_0.$$

Take an inner product in H of this equality and $M(w-u)$ taking into account (7.6), to verify that

$$[M(u-\bar{w}), M(u-w)]_{D^{-1}} = 0. \quad (7.10)$$

2. Methods of gradient descent, gradient method of steepest descent

There is an element u in formulas (7.4)–(7.10) to be determined. We overcome this difficulty as follows. Let the element $u \in H_M$ be a solution to Equation (5.1) for $X = H_M$ and we can calculate rather easily a discrepancy $r = g - Lw$ for $\forall\, w \in H_M$. [We have consciously substituted the notations here; otherwise the operator A would be taken for a selfadjoint operator: we take, instead of Equation (10.6) of Chapter 2 with operator A and right-hand part f, Equation (5.1) with an operator L and right-hand part g]. Then formula (7.4) takes the form

$$\eta_0 = -\mathrm{grad}_D G_M(w) = D^{-1}ML^{-1}r. \quad (7.11)$$

Consequently, we calculate the sequences $u^k \in H_M$ decreasing values of quadrature functionals $G_M(x), \Omega_M(x)$ as follows: knowing u^k, compute

$$r^k = g - Lu^k \quad (7.12)$$

$$\eta_0^k = -\mathrm{grad}_D G_M(u^k) = D^{-1}ML^{-1}r^k \quad (7.13)$$

$$t_0^{k+1} = [\eta_0^k]_D^2/[\eta_0^k]_M^2 = \frac{[ML^{-1}r^k]_{D^{-1}}^2}{[\eta_0^k]_M^2} \quad (7.14)$$

and prescribe for $1 < \omega < 2$

$$u^{k+1} = u^k + \omega(2-\omega)t_0^{k+1}\eta_0^k \tag{7.15}$$

or, which is the same,

$$D(u^{k+1}-u^k) = -\omega(2-\omega)t_0^{k+1}ML^{-1}(Lu^k-g),\ k=0,1,\dots. \tag{7.16}$$

It is the generalized *method of gradient descent*, based on variational principles. Unlike the methods of type (5.4), it is a nonlinear method, since the coefficient t_0^{k+1} here depends nonlinearly on u^k.

For $\omega = 1$, the method (7.12)–(7.16) is called the generalized gradient *method of steepest descent*

$$D(u^{k+1}-u^k) = -t_0^{k+1}ML^{-1}(Lu^k-g). \tag{7.17}$$

For this method, according to (7.10), neighboring discrepancies satisfy the following condition of orthogonality

$$[ML^{-1}r^{k+1}, ML^{-1}r^k]_{D^{-1}} = 0. \tag{7.18}$$

Take these partial cases: the first one gave the name to the gradient methods under consideration, the second was called the *method of minimal discrepancies.*

(a) Let $M = L = L^* > 0$, then

$$\eta_0^k = D^{-1}r^k, \qquad t_0^{k+1} = \frac{(D^{-1}r^k, r^k)}{(LD^{-1}r^k, D^{-1}r^k)} \tag{7.19}$$

and

$$D(u^{k+1}-u^k) = -t_0^{k+1}(Lu^k-g) \tag{7.20}$$

$$[r^{k+1}, r^k]_{D^{-1}} = 0. \tag{7.21}$$

For $D = I$ we have the *classic method of steepest descent* (see Figure 16):

$$\eta_0^k = r^k, \qquad t_0^{k+1} = \frac{(r^k, r^k)}{(Lr^k, r^k)} \tag{7.22}$$

$$u^{k+1} = u^k - t_0^{k+1}(Lu^k-g) \tag{7.23}$$

$$(r^{k+1}, r^k) = 0. \tag{7.24}$$

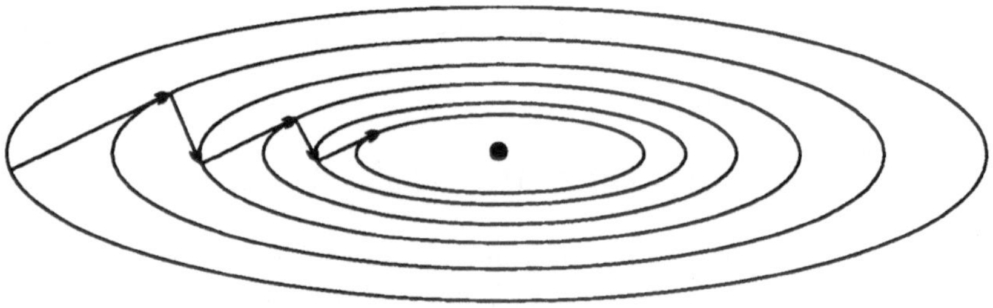

Figure 16

(b) Let $M = L^*BL$ where $B > 0$ is a selfadjoint operator in H. Then

$$\eta_0^k = D^{-1}L^*Br^k, \qquad t_0^{k+1} = \frac{(p^k, D^{-1}p^k)}{(BLp^k, LD^{-1}p^k)}, \tag{7.25}$$

where $p^k = L^*Br^k$ and

$$D(u^{k+1} - u^k) = -t_0^{k+1}L^*B(Lu^k - g) \tag{7.26}$$

$$[L^*Br^{k+1}, L^*Br^k]_{D^{-1}} = 0. \tag{7.27}$$

For $D = B = I$ we get the *classical method of minimal discrepancies* for non-selfadjoint Equation (5.1):

$$\eta_0^k = L^*r^k, \quad t_0^{k+1} = \frac{(L^*r^k, L^*r^k)}{(LL^*r^k, LL^*r^k)} \tag{7.28}$$

and

$$u^{k+1} = u^k - t_0^{k+1}L^*(Lu^k - g) \tag{7.29}$$

$$(L^*r^{k+1}, L^*r^k) = 0. \tag{7.30}$$

If $L^* = L > 0$, then for $D = L$, $B = I$ in (7.25)–(7.27) we obtain formulas of the classic method for selfadjoint Equation (5.1)

$$\eta_0^k = r^k, \qquad t_0^{k+1} = \frac{(Lr^k, r^k)}{(Lr^k, Lr^k)} \tag{7.31}$$

and

$$u^{k+1} = u^k - t_0^{k+1}(Lu^k - g) \tag{7.32}$$

$$(Lr^{k+1}, r^k) = 0. \tag{7.33}$$

We have only to prove that $[u - u^k]_M \to 0$ for $k \to \infty$, that is, that the sequences u^k obtained with formulas (7.17) are minimizing ones for the functional $G_M(x)$.

□ Notice, to this end, that the operator L is a $(D^{-1}ML^{-1}, I)$-symmetrizable operator in H_M and [see (10.15) of Chapter 2)

$$A = D^{-1}M, \qquad f = D^{-1}ML^{-1}g. \tag{7.34}$$

Let $0 < m_A < M_A < \infty$ be lower and upper bounds for the operator A and let

$$\alpha_0 = \frac{2}{m_A + M_A}, \qquad q = \frac{M_A - m_A}{M_A + m_A} < 1. \tag{7.35}$$

Now from the sequence u^k that was obtained with formulas (7.17) we determine a sequence v^k with formulas

$$v^{k+1} = u^k - \alpha_0(Au^k - f) \tag{7.36}$$

or

$$D(v^{k+1} - u^k) = -\alpha_0 ML^{-1}(Lu^k - g). \tag{7.37}$$

Then by virtue of the optimality of the method (7.17) and with regard to the estimate (2.20) in the method (2.12) for $\alpha = \alpha_0$ we obtain that

$$[u - u^{k+1}]_M \leq [u - v^{k+1}]_M \leq q[u - u^k]_M; \tag{7.38}$$

that is,

$$[u - u^{k+1}]_M \leq q^{k+1}[u - u^0]_M. \blacksquare \tag{7.39}$$

Thus, the method (7.17) converges in the space H_M at least as a geometric progression with the denominator q (7.35) to the element u that realizes a minimum for the functional $G_M(x)$.

So, we have examined one-step methods of gradient descent that decrease in an optimal way the error functional $\Omega_M(x)$ too; they demand at each step, as compared to the methods of type (5.4), (5.12), and (5.19), additional arithmetic operations for the computation of coefficients t_0^{k+1}. It would be desirable to obtain the steepest descent for a prescribed number of operations.

There exist traditional methods called the methods of conjugate gradients that for each k realize the optimal minimization of a functional $G_M(x)$ for k steps. For symmetrized equations, they have trinomial form (4.1) (for $u^k = v^k$) where coefficients α_{k+1}, β_{k+1} depend on obtained approximations v^k, v^{k-1}. It is possible, with regard to (7.34), to pass from formulas (7.1) to formulas of type (5.5), (5.13), and (5.20) for approximations u^k as was done in Section 5. The diligent reader, who knows formulas of type α_{k+1}, β_{k+1} for the classical case (7.1), can easily, as we did for the one-step method, obtain formulas for α_{k+1}, β_{k+1} through u^k, u^{k-1} in the generalized methods of conjugate gradients of type (5.5), (5.13), and (5.20) and their partial realizations.

§ 8. Differentiation and Integration of Nonlinear Operators. The Newton Method

Frechét and Gâteaux derivatives. Differentiable functionals and an integral of abstract function. Newton method.

Many problems in applications are essentially nonlinear, which stimulates the development of nonlinear functional analysis. In the following we expound only its original concepts.

Let $\mathbf{X}$ and $\mathbf{Y}$ be Banach spaces, $D \subset \mathbf{X}$ an open set, and let $f(x)$ be an operator mapping D onto $\mathbf{Y}$. An operator (or mapping) $f(x)$ is termed *differentiable in a point $x \in D$ in the Frechét sense* if there exists a bounded linear operator $L_x \in \mathcal{L}(\mathbf{X}, \mathbf{Y})$ such that $\forall x \in \mathbf{X}$ and such that $x + h \in D$,

$$\Delta f = f(x+h) - f(x) = L_x h + \omega(x,h), \tag{8.1}$$

where

$$\|\omega(x,h)\|_{\mathbf{Y}} / \|h\|_{\mathbf{X}} \to 0 \quad \text{for} \quad h \to 0. \tag{8.2}$$

The expression $L_x h$ that is for each $h \in \mathbf{X}$ an element of the space $\mathbf{Y}$ is called the *strong differential* or the *Frechét differential* of the mapping f in the point x and designated by $df(x,h)$, and linear operator L_x is called the *strong derivative* or the *Frechét derivative* of the operator f in the point x and denoted by $f'(x)$. A mapping differentiable in each point of the set D is called the *mapping differentiable on D*. It is easy to verify that a set of mappings determined in a neighborhood of a point $x_0 \in D$, taking values in $\mathbf{Y}$ and differentiable in a point x is a linear system and the differentiation operator is linear; that is,

$$(\alpha f(x) + \beta g(x))' = \alpha f'(x) + \beta g'(x).$$

If $f(x) = y_0 = \text{const}$, then $f'(x) = 0$ and if $f \in \mathcal{L}(\mathbf{X}, \mathbf{Y})$, then for $\forall x \in \mathbf{X}$

$$f'(x) \equiv f \tag{8.3}$$

since in this case $f(x+h) - f(x) = f(h)$. If $\mathbf{Y} = R^1$ is a number line, then f is a functional and $f'(x_0) \in \mathbf{Y}^*$.

Let $h \in \mathbf{X}$, $\|h\| = 1$. A limit

$$\lim_{t \to +0} \frac{f(x+th) - f(x)}{t}, \tag{8.4}$$

if it exists, is called a *derivative of the operator f with respect to direction h* or the *weak* derivative or *Gâteaux derivative* and denoted by $f'_h(x)$.

The Frechét and Gâteaux derivatives are elements of different natures: $f'(x)$ is a linear operator from $\mathcal{L}(\mathbf{X}, \mathbf{Y})$, and $f'_h(x)$ is an element from $\mathbf{Y}$. It is easy to verify that if a mapping f is differentiable in a point x with respect to Frechét, then it is differentiable with respect to Gâteaux for any h and

$$f'_h = f'(x)h. \tag{8.5}$$

The inverse statement is not correct which is exemplified by the following example.

Example 8.1. Let us have a mapping $f : D \to R^m$ where $D \subset R^n$ and R^m, R^n are, respectively, m- and n-dimensional spaces, $n \geq 2$. Then Frechét-differentiability and the Frechét-derivative coincide with the differentiability and derivative from the mathematical analysis for the m-dimensional vector function $f = (f_1, \ldots, f_m)$ of n variables $x = (x_1, \ldots, x_n)$. In this case $f'(x)$ is a linear operator determined by the *Jacobi matrix*:

$$J = \left\{ \frac{\partial f_i}{\partial x_j} \right\}, \qquad i = 1, \ldots, m, \qquad j = 1, \ldots, n\,.$$

The Gâteaux derivative is a derivative of vector function f with respect to direction h; according to (8.5), it is equal to Jh. It is known from mathematical analysis theorems that the existence of a derivative of f in a point x_0 with respect to all directions h does not guarantee the differentiability of f in this point.

For an abstract function $f(x)$ (then $x \in D \subset \mathrm{R}^1$), if limits (8.4) for $h = \pm 1$ differ in a point x just with a sign, then weak differentiability coincides with a strong one and in this case we can regard

$df/dx = f'(x)$ as an element of $\mathbf{Y}$. If the abstract function $f(x)$ is defined on the segment $[a, b]$, then the definite integral of this function

$$\int_a^b f(x)dx$$

is understood as a limit of integral sums

$$\sum_{k=0}^{n-1} f(\xi_k)(t_{k+1} - t_k)$$

corresponding to the partition $0 = t_0 < t_1 < \dots < t_n = b$, $\xi_k \in [t_k, t_{k+1}]$ provided that $\max_k |t_{k+1} - t_k| \to 0$ for $n \to \infty$.

Now let us present the Newton method for the determination of zeros of nonlinear operator equations. More exactly, this method should be named after three authors: Rafson, Newton, and Kantorovich, since Rafson suggested the method before Newton, substituting a nonlinear function with similar linear one (there was no concept of the derivative at that time) and Kantorovich transferred and justified the method for operator equations. This method is also called the method of tangents.

Suppose that there is in D a zero of operator $f(x)$, that is, an element $u \in D$ such that

$$f(u) = 0. \tag{8.6}$$

Assume also that the operator $f(x)$ has in D a continuous Frechét derivative. Take an arbitrary element $u^0 \in D$; if u^0 is close to u, then by (8.1) the element $f(u^0) = f(u^0) - f(u)$ can be substituted with similar expression $f'(u^0)(u^0 - u)$ and, therefore, there is a reason to believe that a solution u^1 of the equation

$$f'(u^0)(u^0 - x) = f(u^0) \tag{8.7}$$

is close to u. Assuming that there exists bounded operator $[f'(u^0)]^{-1}$, we obtain by linear equation (8.7)

$$u^1 = u^0 - [f'(u^0)]^{-1} f(u^0). \tag{8.8}$$

Continuing this process, we obtain the *iterational Newton* method

$$u^{k+1} = u^k - [f'(u^k)]^{-1} f(u^k), \qquad k = 0, 1, \dots . \tag{8.9}$$

It is clear that the Newton method does not always determine a solution of Equation (8.6), since u^k can exceed the boûnds of D, or there can be no $[f'(u^k)]^{-1}$, or the method diverges. Notice that the Newton method for Equation (8.6) coincides with the method of successive approximations (4.3) from Chapter 1 applied to Equation (4.1) from Chapter 1 for

$$P(u) = u - [f'(u)]^{-1} f(u) \tag{8.10}$$

[see also formula (4.12′), Chapter 1].

If the method converges, the operators $f'(u^k)$ and $f'(u^0)$ differ just a little. Therefore, in order to spare the computations, one can propose instead of (8.9) another formula for the method:

$$u^{k+1} = u^k - [f'(u^0)]^{-1} f(u^k) \tag{8.11}$$

that is called the modified Newton method; it corresponds to the transformation of Equation (8.6) into Equation (4.1), Chapter 1 for

$$P(u) = u - [f'(u^0)]^{-1} f(u). \tag{8.12}$$

The convergence condition for the methods (8.9) and (8.11) and the error estimation are presented in the list of monographs on functional analysis.

§ 9. Partial Eigenvalue Problem

Rclay–Temple functional. Two-sided estimates for the least eigenvalue. Iteration method. Error estimate.

Take the following eigenvalue problem in Hilbert space H for selfadjoint operator A:

$$Au = \lambda u. \tag{9.1}$$

According to the assumptions of Section 2 and (2.14) of this chapter, eigenpairs (λ_i, φ_i) will be its solutions. Let $\lambda_1 < \lambda_2 \leq \cdots \leq \lambda_{\max} = M$ and let λ_1 be a simple eigenvalue. A partial problem means the determination of the eigenpair (λ_1, φ_1). A similar problem on the determination of eigenpair $(\lambda_{\max}, \varphi_{\max})$ is reduced to the problem in question by the substitution of the operator $-A$ for the operator A.

Let us know the numbers $\underline{\lambda} \in (\lambda_1, \lambda_2]$ and $\bar{\lambda} \geq M$ and let an element $u \in H$ be assigned with $\|u\| = 1$ and $(\varphi_1, u) > 0$ as the initial approximation for the construction of a sequence $u^k \to \varphi_1$ for $k \to \infty$; it is representable in the form

$$u = \sum_{i=1}^{\infty} a_i \varphi_i = a_i \varphi_i + z, \tag{9.2}$$

where

$$z = \sum_{i=2}^{\infty} a_i \varphi_i, \qquad \sum_{i=1}^{\infty} a_i^2 = 1, \qquad 0 < a_1 \leq 1. \tag{9.3}$$

Obviously, an upper bound for λ_1, according to (9.8) from Chapter 2, is provided by a value of the Relay functional (9.6) of Chapter 2 of $u^k : \lambda_1 \leq \Phi_A(u^k)$. Temple has noticed that the knowledge of the value $\underline{\lambda}$ makes it possible to construct a new functional $\tilde{\Phi}_Q(u)$ that provides for u^k close enough to φ_1 a lower bound: $\tilde{\Phi}_Q(u^k) \leq \lambda_1$. Let us construct it as follows. Let real bounded function $Q(\lambda)$ be assigned on the set $\lambda_1 \cup [\underline{\lambda}, \bar{\lambda}]$ and let $Q(\lambda_1) > 0$. Denote

$$\omega = \omega(Q) = \max_{\lambda \in [\underline{\lambda}, \bar{\lambda}]} |Q(\lambda)|/Q(\lambda_1) \tag{9.4}$$

and define the *Relay–Temple functional* with the formula

$$\tilde{\Phi}_Q(u) = \frac{(AQ(A)u, u)}{(Q(A)u, u)}. \tag{9.5}$$

Its value on an element u of the form (9.2) equals

$$\tilde{\Phi}_Q(u) = \frac{\sum\limits_{i=1}^{\infty} \lambda_i Q(\lambda_i) a_i^2}{\sum\limits_{i=1}^{\infty} Q(\lambda_i) a_i^2}. \tag{9.6}$$

It is obvious that $\tilde{\Phi}_Q(\varphi_1) = \lambda_1$. Let us use this functional for the approximate calculation of λ_1 in the nontrivial case: for $u \neq \varphi_1$, $0 < |a_1| < 1$, and $(Q(A)z, z) \neq 0$ if either $Q(\lambda) \geq 0$, or $Q(\lambda) \leq 0$ for $\lambda \in [\underline{\lambda}, \bar{\lambda}]$. Then

$$\lambda_2 \leq \tilde{\Phi}_Q(z) = \Phi_A(|Q(A)|^{1/2} z) \leq M. \tag{9.7}$$

Let

$$a_1^{-2} - 1 = \text{tg}^2 \psi, \qquad B = \frac{\sum\limits_{i=1}^{\infty} Q(\lambda_i) a_i^2}{Q(\lambda_1) a_1^2} \tag{9.8}$$

$$C = \frac{\sum_{i=2}^{\infty}(\lambda_i - \lambda_1)Q(\lambda_i)a_i^2}{Q(\lambda_1)a_1^2(1+B)}. \tag{9.9}$$

Simple calculations demonstrate that

$$\tilde{\Phi}_Q(u) - \lambda_1 = C \tag{9.10}$$

$$\tilde{\Phi}_Q(z) - \tilde{\Phi}_Q(u) = C/B; \tag{9.11}$$

that is,

$$\Delta(u) = \frac{\tilde{\Phi}_Q(u) - \lambda_1}{\tilde{\Phi}_Q(z) - \tilde{\Phi}_Q(u)} = B. \tag{9.12}$$

Let

$$\delta(u) = \frac{\tilde{\Phi}_Q(u) - \lambda_1}{M - \tilde{\Phi}_Q(u)}. \tag{9.13}$$

The following theorem determines two-sided estimates for λ_1.

Theorem 1.

(1) *If* $Q(\lambda) \geq 0$ *for* $\lambda \in [\underline{\lambda}, \bar{\lambda}]$, *then*

$$0 \leq \delta(u) \leq \omega \operatorname{tg}^2 \psi \tag{9.14}$$

(2) *If* $Q(\lambda) \leq 0$ *for* $\lambda \in [\underline{\lambda}, \bar{\lambda}]$ *and*

$$\omega \operatorname{tg}^2 \psi < 1, \tag{9.15}$$

then

$$0 \leq -\delta(u) \leq \omega \operatorname{tg}^2 \psi. \tag{9.16}$$

□ Use (9.4) and (9.8) to obtain $|B| \leq \omega \operatorname{tg}^2 \psi$. Consequently, by (9.8) and (9.9) we have that $B \geq 0$, $C \geq 0$ in the first case and $-1 < B \leq 0$, $C \leq 0$ in the second. Then by (9.7) and (9.11)

$$M - \tilde{\Phi}_Q(u) \geq \tilde{\Phi}_Q(z) - \tilde{\Phi}_Q(u) = C/B \geq 0. \tag{9.17}$$

Therefore, the left inequalities in (9.14) and (9.16) follow from formulas (9.10) and (9.17). The upper bound in (9.14) follows from (9.10), (9.12), and (9.17):

$$\delta(u) \leq \Delta(u) = B \leq \omega \operatorname{tg}^2 \psi;$$

similarly, in the second case

$$-\delta(u) \le -\Delta(u) = -B \le \omega\,\mathrm{tg}^2\psi. \blacksquare$$

Take, as the functions $Q(\lambda)$, two classes of power N polynomials $Q_N^{\pm}(\lambda)$ such that $Q_N^{\pm}(\lambda_1) > 0$ and $Q_N^{+}(\lambda) \ge 0$ and $Q_N^{-}(\lambda) \le 0$ for $\lambda \in [\underline{\lambda}, \bar{\lambda}]$. Solve on these classes a minimization problem for the values $\omega(Q_N^{+}(\lambda))$ and $\omega(Q_N^{-}(\lambda))$. A solution for these problems follows from Corollary 2, Section 10 of Chapter 1 according to which

$$\min_{Q_N^{+}} \omega(Q_N^{+}) = \omega(R_N + 1) = 2/(T_N(\theta) + 1) \tag{9.18}$$

and

$$\min_{Q_N^{-}} \omega(Q_N^{-}) = \omega(R_N - 1) = 2/(T_N(\theta) - 1), \tag{9.19}$$

where

$$R_N = R_N(\lambda) = T_N\left(\frac{\bar{\lambda} + \underline{\lambda} - 2\lambda}{\bar{\lambda} - \underline{\lambda}}\right) \tag{9.20}$$

$$\theta = (\bar{\lambda} + \underline{\lambda} - 2\lambda_1)/(\bar{\lambda} - \underline{\lambda}) > 1. \tag{9.21}$$

So, for extremal polynomials $R_N \pm 1 : \omega \to 0$ for $N \to \infty$ and for sufficiently large N inequality (9.15) is satisfied. We suggest the optimal [in the sense of minimization of the $\omega(Q_N^{\pm})$] iteration method to find the eigenpair (λ_1, φ_1) that realizes the estimates (9.18) and (9.19).

Let $N = 2n$, then (see Section 10 of Chapter 1)

$$T_N(x) + 1 = 2T_n^2(x);$$

that is,

$$Q_N^{+}(\lambda) = R_N(\lambda) + 1 = 2R_n^2(\lambda)$$

and the functional $\tilde{\Phi}_{Q_N^{+}}(u)$ can be written in the form of the Relay functional:

$$\tilde{\Phi}_{Q_N^{+}}(u) = \frac{(AR_n(A)u, R_n(A)u)}{(R_n(A)u, R_n(A)u)} = \Phi_A(R_n(A)u). \tag{9.22}$$

Apply, in order to find its value equal to $\lambda_n^{+} \ge \lambda_1$, the Chebyshev iteration method: construct starting with an assigned value $u^0 = u$ a sequence u^k according to the algorithm:

$$\begin{aligned} v^{k+1} &= Au^k - \gamma_{k+1}u^k, \\ u^{k+1} &= v^{k+1}/\|v^{k+1}\|, \\ k &= 0, 1, \ldots, n-1, \end{aligned} \tag{9.23}$$

where

$$\gamma_j = \frac{1}{2}(\bar{\lambda} + \underline{\lambda} - (\bar{\lambda} - \underline{\lambda})\beta_j) \tag{9.24}$$

and β_j are ordered roots $T_n(x)$ (see Section 3). Then

$$u^n = R_n(A)u^0 / \|R_n(A)u^0\| \tag{9.25}$$

and according to (9.23)

$$\|R_n(A)u^0\| = \prod_{i=1}^{n} \|v^i\|. \tag{9.26}$$

If $z^n = Au^n$, then by (9.22)

$$\lambda_n^+ = (z^n, u^n) \geq \lambda_1. \tag{9.27}$$

We use, in order to calculate a value of $\tilde{\Phi}_{Q_N^-}(u)$ for $Q_N^-(\lambda) = R_N(\lambda) - 1$ equal to $\lambda_n^- \leq \lambda_1$, the same results of method (9.23), since

$$\Phi_{Q_N^-}(u) = \frac{(A(R_N(A)+I)u, u) - 2(Au, u)}{((R_N+I)u, u) - 2(u, u)}$$

$$= \frac{(AR_n u, R_n u) - (Au, u)}{(R_n u, R_n u) - (u, u)}. \tag{9.28}$$

Substitute $u = u^0$ and $R_n(A)u^0$ from (9.25) into (9.28) to obtain

$$\lambda_n^- = \frac{(z^n, u^n)\|R_n(A)u^0\|^2 - (Au^0, u^0)}{\|R_n(A)u^0\|^2 - 1}, \tag{9.29}$$

where $\|R_n(A)u^0\|$ is determined in (9.26). So, for $n \geq n_0 = n_0(a_1, \theta)$

$$\lambda_n^- \leq \lambda_1 \leq \lambda_n^+. \tag{9.30}$$

Now let us estimate now an error $\varepsilon^n = \varphi_1 - u^n$ assuming, with no loss of generality, that $0 < a_1 < 1$:

$$\|\varepsilon^n\|^2 = (\varphi_1 - u^n, \varphi_1 - u^n) = \left(1 - \frac{R_n(\lambda_1)a_1}{\|R_n(A)u^0\|}\right)^2 + \frac{\sum_{i=2}^{\infty} R_n^2(\lambda_i)a_i^2}{\|R_n(A)u^0\|^2}$$

$$= \|R_n(A)u^0\|^{-2}[(\|R_n(A)u^0\| - R_n(\lambda_1)a_1)^2 + \sum_{i=2}^{\infty} R_n^2(\lambda_i)a_i^2].$$

Transform the expression in the first term in square brackets:

$$\|R_n(A)u^0\| - R_n(\lambda_1)a_1 = \frac{\|R_n(A)u^0\|^2 - R_n^2(\lambda_1)a_1^2}{\|R_n(A)u^0\| + R_n(\lambda_1)a_1}$$

$$= \frac{\sum\limits_{i=2}^{\infty} R_n^2(\lambda_i)a_i^2}{\|R_n(A)u^0\| + R_n(\lambda_1)a_1}.$$

Consequently,

$$\|\varepsilon^n\|^2 = \frac{\sum\limits_{i=2}^{\infty} R_n^2(\lambda_i)a_i^2}{\|R_n(A)u^0\|^2}\left(1 + \frac{\sum\limits_{i=2}^{n} R_n^2(\lambda_i)a_i^2}{(\|R_n(A)u^0\| + R_n(\lambda_1)a_1)^2}\right).$$

Taking into account that $R_n^2(\lambda_i) \leq 1$ for $i \geq 2$ and $R_n(\lambda_1) = T_n(\theta)$ where θ is determined in (9.21), we obtain an estimate for $\|\varepsilon^n\|$:

$$\|\varepsilon^n\|^2 \leq \frac{\mathrm{tg}^2\psi}{T_n^2(\theta)}\left(1 + \frac{\mathrm{tg}^2\psi}{4T_n^2(\theta)}\right). \qquad (9.31)$$

It is evident that the method (9.23)–(9.27) and (9.29) can be considered as infinitely continyable where n takes the values 2^p3^r, $r = 0, 1, \ldots$, and the values β_j in (9.24) are formed on the bases of the T-sequence (see Section 3 of this chapter).

The partial problem for a generalized eigenvalue problem

$$Lv = \lambda Mv, \qquad (9.32)$$

where $L = L^* > 0$, $M = M^* > 0$ can be reduced, in some cases, with substitution and transformation of Equation (9.32) to problem (9.1). For example, if $M = B^*B$, then with the substitution $Bv = u$ or $M^{-1/2}v = u$ and miltiplying (9.32) from the left by B^{*-1} (or by $M^{-1/2}$) we transform Equation (9.32) into (9.1) with $A = B^{*-1}LB^{-1}$ ($A = M^{-1/2}LM^{-1/2}$). Then (see Section 5) it is possible to formulate the iteartion method for a transformed equation through inverse substitution as an iteration method for initial equation (9.32).

Remark. We have discussed in Sections 3, 4, and 9 of this chapter iteration methods for problems with selfadjoint, positive definite opeartors; the same methods are applicable to problems with bounded operators that have a positive real spectrum.

We reduced the initial equation (complex spectrum is possible) in Sections 5 and 7 of this chapter with (C, D)-symmetrization into an equation with a selfadjoint operator; such transformation is not always feasible. However, if it is known that the complex spectrum of a problem lies inside a lemniscate of a polynomial $Q_m(z)$ embracing this spectrum densely enough, say, of the type (11.57), Chapter 1, then it is possible, by Theorem 1, Section 11 of Chapter 1, to apply for the solution of such problems the methods of type (3.1) where a transition operator $P_{ml}(A)$ is determined by formula (11.55) of Chapter 1 and the parameters α_{k+1} are equal to inverse values of the roots of the polynomial $P_{ml}(z)$ (see Examples 11.1, 11.2 of Chapter 1).

§ 10. Successive Approximation Method for Inverse Operator

On problems effectively solved with approximations to inverse operator. Method for obtaining finite sums of the Neumann series.

As we elucidated before, it is possible, if we know the inverse operator A^{-1} to an operator A, to express through it a solution of an operator equation of the first kind [see formulas (3.4) and (3.5), Chapter 2], and if we also know the operators H_k that are sufficiently close to A^{-1}, then with conditions (2.10) being satisfied, to construct fast converging linear one-step iteration methods of type (1.20). Various modifications of the Newton method imply the realization of a multiplication operation for $A^{-1} = (f'(u))^{-1}$ or similar operator by some element.

Consider, finally, one more important problem: the refinement of a solution to Equation (1.17) obtained in real calculations. Let $\tilde{u}$ be an approximate solution of this equation [$\tilde{u}$ may be understood also as a "solution" calculated by computer with some theoretically exact (noniteration) method, since in real computations, because of finite digit length and roundoff errors, we practically always get an approximate result]. Then, if $r = f - A\tilde{u}$ is a discrepancy, an error $\varepsilon = u - \tilde{u}$ satisfies (see (3.7), Chapter 2) the equation

$$A\varepsilon = r. \tag{10.1}$$

Therefore, if the operator H_k is close or equal to A^{-1}, then one may expect that a major part of the error $\tilde{\varepsilon}$ is represented by the formula

$$\tilde{\varepsilon} = H_k r \tag{10.2}$$

and, consequently, the element

$$\tilde{u} + H_k r = \tilde{u} - H_k(A\tilde{u} - f) \tag{10.2}$$

is closer to an exact solution. This refinement may be repeated in real computations using various operators H_k and arriving thus at the iteration method (1.21).

In the following we set off one of the methods for the construction of the operator sequence H_k that converges fast enough to A^{-1}. This method is a generalization of the iteration formula for the computation of inverse numbers, obtained with the Newton method; it is a rather efficient algorithm for the computation of partial sums of the Neumann series (4.7) of Chapter 2.

So, let the operators $A, A^{-1}, H, H^{-1} \in \mathcal{L}(X)$. Determine a sequence of operators $H_k, B_k \in \mathcal{L}(X)$ with the formulas

$$B_k = I - H_k A, \qquad H_{k+1} = (I + B_k)H_k, \tag{10.3}$$

where

$$H_0 = H, \qquad B_0 = I - HA. \tag{10.4}$$

As $H_k A = I - B_k$, then

$$B_{k+1} = I - (I + B_k)H_k A = I - (I + B_k)(I - B_k) = B_k^2; \tag{10.5}$$

that is,

$$B_k = B_0^{2^k}. \tag{10.6}$$

Let

$$\rho = \|I - HA\| < 1. \tag{10.7}$$

For example, if $A = A^* > 0$ is a selfadjoint positive definite operator and m_A and M_A are its bounds (see Section 8 of Chapter 2), then inequality (10.7) with $\rho = (M_A - m_A)/(M_A + m_A)$ will be satisfied if $H = \alpha I$ is taken for $\alpha = 2/(M_A + m_A)$.

With condition (10.7) being satisfied, by (10.6) $\|B_k\| \leq \rho^{2^k}$ and the products

$$V_k = (I + B_k) \cdot (I + B_{k-1}) \cdots (I + B_0) = \sum_{i=0}^{2^{k+1}-1} B_0^i$$

are finite sums of the convergent Neumann series whose sum, according to (4.12) of Chapter 2, equals

$$(I - B_0)^{-1} = A^{-1}H^{-1}.$$

Consequently, the sequence of operators

$$H_{k+1} = (I + B_k) \cdot (I + B_{k-1}) \cdots (I + B_0)H \tag{10.8}$$

converges with respect to the norm to the operator A^{-1}; more than that, as

$$H_{k+1} - A^{-1} = - \sum_{i=2^{k+1}}^{\infty} B_0^i H \tag{10.9}$$

then, with regard to (10.4) and (10.7) and estimating a residual of series (10.9), we obtain

$$\|H_{k+1} - A^{-1}\| \leq \|H\| \rho^{2^{k+1}}/(1 - \rho). \tag{10.10}$$

As a result of the preceding analysis, the method for the search for the sequence H_k can be written in the form

$$H_{k+1} = (I + B_k)H_k, \qquad B_{k+1} = B_k \cdot B_k, \qquad k = 0, 1, \ldots \tag{10.11}$$

for initial conditions (10.4).

Everything implicit (secret)
becomes explicit.

§ 11. Stability and Optimization of Explicit Difference Schemes for Stiff Differential Equations

Initial value problem, explicit and implicit difference schemes, assumptions, model stiff problem, courant. Approximation and stability conditions. Optimization of parameters for explicit difference schemes.

1. Stiff initial value problem, assumptions. Explicit and implicit difference schemes, courant

Take an initial value problem for $t_0 \leq t \leq T$ for a differential equation in Banach space $\mathbf{X}$

$$\frac{du}{dt} = f(u, t) \tag{11.1}$$

$$u|_{t=t_0} = u_0, \tag{11.2}$$

where $f(u,t)$ is a nonlinear operator $\mathbf{X} \to \mathbf{X}$ of two variables: real variable $t \geq t_0$ (time) and $u = u(t) \in \mathbf{X}$, function of t. The contraction mapping principle (see Section 4 of Chapter 1) applied to integral equation (4.32) of Chapter 1 equivalent to problem (11.1) and (11.2) permits, under certain requirements from $f(u,t)$, formulation of the existence and uniqueness conditions for a solution to problem (11.1) and (11.2). We assume here that these conditions are met; $f(u(t),t)$ and $u(t)$ are sufficiently smooth with respect to t functions for (11.1) to be substituted with a difference equation with prescribed order of local approximation. More exactly, for given $\varepsilon > 0$ (accuracy of local approximation) let $\tau > 0$ ($\tau < T - t_0$) be known such that for $|\Delta t| \leq \tau$

$$\frac{du}{dt} = \frac{u(t+\Delta t) - u(t)}{\Delta t} + \eta(t), \tag{11.3}$$

where

$$\|\eta(t)\| \leq \varepsilon, \quad \text{for} \quad t_0 \leq t, \qquad t + \Delta t \leq T. \tag{11.4}$$

Let $t_0 = 0 < t_1 < t_2 < \cdots < t_k < t_{k+1} < \cdots \leq T$; $\tau_{k+1} = t_{k+1} - t_k$, $\tau_{k+1} \leq \tau$ are time steps and let $u_k = u(t_k)$. For the sake of simplicity, we use the same notation for approximate solutions. Let $t_k \leq t \leq t_{k+1}$. If we set $t = t_{k+1}$ in Equation (11.1) and substitute the derivative du/dt, neglecting $\eta(t)$, with formula (11.3) for $\Delta t = -\tau_{k+1}$, then we arrive at the formulas for the implicit Euler method:

$$u_{k+1} = u_k + \tau_{k+1} f(u_{k+1}, t_{k+1}), \qquad k = 0, 1, \dots . \tag{11.5}$$

If we assign in (11.1) and (11.3) $t = t_k$, $\Delta t = \tau_{k+1}$, then the formulas of the explicit Euler method are obtained:

$$u_{k+1} = u_k + \tau_{k+1} f(u_k, t_k). \tag{11.6}$$

The implicit method (11.5) implies for the determination of u_{k+1} the solution (with the Newton method, e.g.) nonlinear equation (11.5). Explicit method is free of this deficiency; however, stable realization of this method may result in strong restrictions for time steps.

We analyze the problems that appear in the realization of explicit methods using the model initial value problem to be obtained after the following series of simplifications that are assumed to be justified for the approximate solution of the original initial value problem (11.1) and (11.2) in a neighborhood of (u_0, t_0).

Let an operator $f(u,t)$ be Frechét-differentiable and let $J(u,t)$ be its Frechét derivative. Substitute (11.1) with the linear differential equation

$$\frac{du}{dt} = J(u_0, t)(u - u_0) + f(u_0, t)$$

that substitutes in turn with the equation with constant operator at u:

$$\frac{du}{dt} = -Au + b(t), \tag{11.7}$$

where $A = -J(u_0, t_0) \in \mathcal{L}(X)$, $b(t) = f(u_0, t) - J(u_0, t)u_0 \in \mathbf{X}$.

Now let us take, instead of the approximate initial value problem (11.7) and (11.2), a linear ordinary initial value problem in the form

$$\frac{du}{dt} = -Au \tag{11.8}$$

$$u|_{t=0} = u_0 \tag{11.9}$$

for $0 \leq t \leq T$, since major problems, connected with the approximation and stability while employing explicit difference schemes, appear while solving such problems too.

Let the operator A have a just nonnegative discrete spectrum, (λ_i, φ_i) be eigenpairs of the operator A, and let $\{\varphi_i\}$ form a basis in space $\mathbf{X}$, $M = \sup_i \lambda_i < \infty$.

Let

$$u_0 = \sum_i a_i \varphi_i \tag{11.10}$$

then a solution of problem (11.8) and (11.9) has a form

$$u(t) = \sum_i \exp(-\lambda_i t) a_i \varphi_i. \tag{11.11}$$

Notice that a norm of each of the terms in (11.11) does not increase for $t \to \infty$.

Let us call an initial value problem a *stiff initial value problem* if

$$TM \gg 1. \tag{11.12}$$

Formulas of the *implicit method* (11.5) take for problem (11.8) and (11.9) the form

$$(I + \tau_{k+1} A) u_{k+1} = u_k$$

or

$$u_{k+1} = (I + \tau_{k+1} A)^{-1} u_k \tag{11.13}$$

and for the *explicit method*

$$u_{k+1} = (I - \tau_{k+1}A)u_k. \tag{11.14}$$

It was required, naturally, with regard to the form of solution (11.11) to problem (11.8) and (11.9), that spectral radii of transition operators in (11.13) and (11.14) for constant step $\tau_{k+1} = h$ would not exceed the unit; otherwise we would not only lose the approximation property, but would encounter catastrophic growth of norms of approximate solutions. For method (11.13) we have, using formula (4.5) of Chapter 2: $\mu((I + hA)^{-1}) \leq 1$ and for method (11.14) we have $\mu(I - hA) \leq \max\{|1 - hM|, 1\}$. Consequently, if $h \leq \text{cou}$ where

$$\text{cou} = 2/M,$$

then $\mu(I - hA) \leq 1$. Let us call the value cou the *courant* after R. Courant who investigated properties of explicit difference schemes.

So, if all $\tau_{k+1} = h$, then the value h in the explicit Euler method

$$u_{k+1} = u_k - hAu_k \tag{11.15}$$

or for general Equation (11.16)

$$u_{k+1} = u_k + hf(u_k, t) \tag{11.16}$$

must satisfy the inequality

$$h \leq \text{cou}. \tag{11.17}$$

We see, comparing inequalities (11.12) and (11.17), that they demand a lot of time steps to solve an initial value problem. A justified question arises: do there exist explicit stable algorithms with variable steps τ_{k+1} that integrate stiff initial value problems with an essentially smaller number of steps?

2. Approximation and stability conditions

Let $N \geq 1$ be an integer; perform a cycle of N steps with formulas (11.14), starting with $t_0 = 0$. Then we have

$$u_N = P_N(A)u_0, \tag{11.18}$$

where

$$P_N(\lambda) = \prod_{i=1}^{N}(1 - \tau_i\lambda). \tag{11.19}$$

Then

$$P_N(0) = 1 \quad \text{and} \quad l_N = \sum_{i=1}^{N} \tau_i = -P_N'(0) \tag{11.20}$$

and τ_i are the values inverse to the roots of $P_N(\lambda)$. As

$$\exp(-l_N\lambda) = 1 - l_N\lambda + \frac{l_N^2\lambda^2}{2} - \dots \tag{11.21}$$

and

$$P_N(\lambda) = 1 - l_N\lambda + \sum_{i,k} \tau_i\tau_k\lambda^2 + \dots \tag{11.22}$$

then, comparing (11.21) and (11.22) we see that

$$|\exp(-l_N\lambda) - P_N(\lambda)| = 0(\lambda^2); \tag{11.23}$$

that is, the approximation of the operator $\exp(-At)$ by the operator $P_N(A)$ is performed.

So, let a size of maximal time step τ be determined for chosen accuracy of local approximation $\varepsilon > 0$ by conditions (11.3) and (11.4) (the dependence of an error between exact and difference solutions on ε presents a more complicated problem; we do not discuss it here).

It is natural to demand that the following restrictions and conditions be satisfied which provide for a small value of an error of local approximation in (11.4) and (11.23) for the cycle of N steps.

(1)
$$\max_{1 \le k \le N} \tau_k = \tau \tag{11.24}$$

(2)
$$\tau \le l_N \le B_0\tau \tag{11.25}$$

where $B_0 > 1$ is a value independent of N;

(3) conditions of spectral stability:

$$\max_{0 \le \lambda \le M} |P_N(\lambda)| \le 1 \tag{11.26}$$

since in this case

$$\mu(P_N(A)) = \sup_{\lambda \in S_P(A)} |P_N(\lambda)| \le 1; \tag{11.27}$$

(4) conditions of stable realization of computations inside the cycle of N steps.

3. Parameter optimization for explicit difference schemes

The following values are useful for the comparison of method (1.14) to method (11.15) for $h = \text{cou}$:

$$g = \tau/\text{cou}, \qquad V_N = \frac{l_N}{N\,\text{cou}}. \tag{11.28}$$

The first one shows how many times τ is larger than cou, and the second is equal to a ratio between the lengths of integration intervals for N steps of methods (11.14) and (11.15) for maximal step $h = \text{cou}$.

Let us consider and investigate the possibilities of normed by the first condition (11.20) and reduced to the segment $[m, M]$ $(m \geq 0)$ Chebyshev polynomials of the first kind (3.10), for which m and N are the parameters to be determined, as a class of polynomials that satisfies the condition (11.26).

We see that such a polynomial $P_N(\lambda)$ decreases on the segment $[0, m]$ and

$$\max_{m \leq \lambda \leq M} |P_N(\lambda)| = \frac{1}{|T_N(\Theta)|} = \eta_N \leq 1, \tag{11.29}$$

where Θ is determined in (3.9). Using formulas (3.10), (3.8), and (10.20) of Chapter 1, compute for $\lambda = 0$ a derivative of $P_N(\lambda)$, and, consequently, by (11.20) the value l_N too:

$$l_N = -P_N'(0) = \frac{2NU_{N-1}(\Theta)}{(M-m)T_N(\Theta)}. \tag{11.30}$$

According to Theorem 5 from Section 10 of Chapter 1, all other polynomials $Q_N(\lambda)$ of Nth power satisfying the condition $|Q_N(\lambda)| \leq \eta_N$ for $\lambda \in [m, M]$ have smaller in a module derivative for $\lambda = 0$. Transform, using formulas (3.9) and (3.12) for Θ and σ and formulas (10.8) and (10.18) of Chapter 1 for $T_N(\Theta)$ and $U_{N-1}(\Theta)$ the expression (11.30) to the form

$$l_N = V_N N\,\text{cou}, \tag{11.31}$$

where [see (11.28)]

$$V_N = V_N(\sigma) = \frac{1}{2}\frac{1+\sigma}{1-\sigma} \cdot \frac{1-\sigma^{2N}}{1+\sigma^{2N}} \tag{11.32}$$

and $0 \le \sigma \le 1$. The function $V_N(\sigma)$ is a monotonically increasing function of σ, $V_N(1) = N$ and a monotonically increasing function of N. For the extremal case, when $\sigma \to 1$ $(m \to 0)$, we have

$$V_N = V_N(1) = N, \qquad l_N = N^2 \text{cou}; \tag{11.33}$$

that is, using the Chebyshev polynomial, we can by N steps in time go N times farther as compared to the explicit Euler method (11.15) with constant, maximal permissible step $h = \text{cou}$.

It is obvious that if $g \le 1$, then one should take $N = 1$, $\tau_1 = \tau$; in this case, the steps are determined both in explicit and implicit schemes just by local approximation and must be equal. Let $N > 1$; introduce the following notations: $a = \Theta^{-1}$ $(0 \le a \le 1)$, $h^0 = (1 + a)\text{cou}/2$, $\beta_0 = \cos \pi/2N$, $\kappa_N = (j_1, \dots, j_N)$, and $\beta_k = \cos \omega_k \pi$, $k = 1, \dots, N$, where $\omega_k = (2j_k - 1)/(2N)$. Then by (3.9) and (3.17) we obtain the formulas for time steps ·

$$\tau_k = h^0/(1 - a\beta_k), \qquad k = 1, \dots, N. \tag{11.34}$$

Moreover,

$$\tau_0 = \max_k \tau_k = h^0/(1 - a\beta_0). \tag{11.35}$$

Now determine N and m. Determine first the minimal N, for which in a limit case $\sigma = 1$ (i.e., $a = 1$) maximal step τ_0 (11.35) is not larger than τ. Assign in (11.35) $a = 1$, $\tau_0 = \tau$ to obtain [see (11.28)]

$$g = \frac{1}{2\sin^2 \frac{\pi}{4N}}. \tag{11.36}$$

From this relationship, we obtain the following formula for N:

$$N = \left[\frac{\pi}{4} / \arcsin(1/\sqrt{2g})\right] + 1, \tag{11.37}$$

where $[z]$ denotes an integer part of z. We see thus, that for $g \gg 1$

$$N^2 \approx \frac{\pi^2}{8} g,$$

and [see (11.28) and (11.33)]

$$l_N \approx \frac{\pi^2}{8} \tau. \tag{11.38}$$

That is, for our choice of N we have proven by formula (11.37), the existence of the constant B_0 in inequality (11.25).

Determine the value a in (11.34) and, consequently, the value m too, from conditions (11.24); we have with regard to (11.35)

$$a = \frac{2g-1}{2g\beta_0+1}. \tag{11.39}$$

We have only the question of stable realization of the method to be considered. Comparing formulas of the Chebyshev method (3.1) for $f = 0$ and (3.17) with the method (11.14) and (11.34), we see that $u^k = u_k$ for $u^0 = u_0$, $\alpha_k = \tau_k$, $k = 1, \dots, N$. Consequently, the stability analysis performed in Section 3 of this chapter, its criteria for values (3.20), and the algorithm for mixing the parameters are appropriate for the analysis of stability and for the determination of an order of steps (11.34) in the explicit method (11.14). So, we determine the time steps (11.34) with the permutation κ_N that guarantees the stability. Having performed a cycle of N steps, determine the new value of τ and determine with it a new cycle length, and so on.

Remark. It is possible to show that condition (11.26) will be satisfied for the preceding Chebyshev polynomial (3.10) and for problems (11.8) with complex spectrum $Sp(A)$ if it lies inside a line of the level Ω_ρ with $\rho = 1$ (see Section 11 of Chapter 1).

The method we have suggested presumes that the roots of polynomial (11.19) are real; high dispersion of the values $\{\tau_i\}_1^N$ [see (11.38)] is observed here for $g \gg 1$. Therefore, let us obtain another formulas for the explicit method. Let, for the sake of simplicity, $N = 2n$; then the polynomial $P_N(\lambda)$ (11.19) can be factorized into quadrature multipliers using the fact that

$$(1-\tau_i\lambda)(1-\tau_{i+1}\lambda) = (1-h_i\lambda)^2 - \gamma_i h_i^2\lambda^2, \tag{11.40}$$

where

$$h_i = \frac{1}{2}(\tau_i + \tau_{i+1}), \quad \gamma_i = \left(\frac{\tau_{i+1}-\tau_i}{\tau_{i+1}+\tau_i}\right)^2. \tag{11.41}$$

Substitute (11.40) into (11.19) to obtain

$$P_N(\lambda) = \prod_{i=1}^{n}[(1-h_i\lambda)^2 - \gamma_i h_i^2\lambda^2]. \tag{11.42}$$

We assume here that if τ_i is complex, then $\tau_{i+1} = \bar{\tau}_i$ and if τ_i is determined with formula (11.34), then τ_i and τ_{i+1} form a pair $h^0/(1 \pm a\beta_k)$; in the latter case,

$$h_i = h^0/(1 - a^2\beta_i^2), \qquad \gamma_i = a^2\beta_i^2. \tag{11.43}$$

Then two steps of method (11.14) can be realized with formulas

$$\begin{aligned} y_{k+1/2} &= u_k - h_{k+1}Au^k \\ y_{k+1} &= y_{k+1/2} - h_{k+1}Ay_{k+1/2} \\ y_{k+2} &= y_{k+1} + \gamma_{k+1}h_{k+1}(Ay_{k+1/2} - Au^k). \end{aligned}$$

These formulas permit the construction of the following formulas for the integration of problem (11.1) and (11.2)

$$\begin{aligned} u_{k+1/2} &= u_k + h_{k+1}f(u_k, t_k) \\ t_{k+1/2} &= t_k + h_{k+1} \\ u_{k+1} &= u_{k+1/2} + h_{k+1}f(y_{k+1/2}, t_{k+1/2}) \\ t_{k+1} &= t_{k+1/2} + h_{k+1} \\ u_{k+1} &= y_{k+1} + \gamma_{k+1}h_{k+1}(f(u_k, t_k) - f(y_{k+1/2}, t_{k+1/2})) \end{aligned} \tag{11.44}$$

$$k = 0, 1, \ldots, n-1$$

that consist of two steps by the Euler method and one refinement.

In this method, if h_i are determined with formulas (11.43), for $g \gg 1$ the maximal step is approximately by two times smaller and minimal by two times larger as compared to the preceding method. For problems with a complex spectrum, the polynomials of the type (11.42) may be constructed as polynomials from (11.57) of Chapter 1. Then their roots are found explicitly, and their parameters should be selected so that the stability range reflects the spectrum characteristics of the problem in question (see Examples 11.1 and 11.2 in Section 11 of Chapter 1).

Index

Bibliography

Mathematical Analysis:

1. Kudryavtsev L.D. *Mathematical Analysis.* Vol. 1, 2. M.: Vysshaya shkola, 1981 (in Russian).
2. Bary I.L. *Trigonometrical Series.* M.: Phyzmatgiz, 1961 (in Russian).
3. Hairer E., Wanner G. *Analysis by Its History.* Undergraduate Texts in Mathematics. New-York, Berlin, Heidelberg: Springer–Verlag, 1995.

Linear Algebra:

4. Faddeev D.K., Faddeeva V.N. *Computational Methods of Linear Algebra.* Frecman W.H & Co., San Francisco, 1963.

Theory of Approximations:

5. Natanson I.P. *Constructive Theory of Functions* M.–L.: Gos. izdat. tecn.-teor. lit., 1949 (in Russian).
6. Smirnov V.I, Lebedev N.A. *Constructive Theory of Complex Variable.* M.: Nauka, 1964 (in Russian).
7. Tikhomirov V.M. Some Aspects of the Approximation Theory. M.: Izd. MGU, 1976 (in Russian).
8. Pashkovski C. *Computational Applications of the Chebyshev Polynomials and Series.* - M.: Nauka, 1982 (in Russian).
9. Krein M.G., Nudelman A.A. *Problems of the Markov Moments and Extremal Problems.* - M.: Nauka, 1973 (in Russian).

Equations of Mathematical Physics and Solution Methods:

10. Vladimirov V.C. *Equations of Mathematical Physics.* M.: Nauka, 1971 (in Russian).
11. Mikhailov V.P.*Partial Differential Equations.* M.: Nauka, 1976 (in Russian).
12. Mikhlin S.G. Mathematical Physics. M.: Nauka, 1968 (in Russian).
13. Marchuk G.I, Lebedev V.I. *Numerical Methods in the Theory of Neutron Transport.* London, Paris, New York: Harwood Academic Publishers 1986.

14. Aubin J. *Approximation of Elliptic Boundary Value Problems.* Wiley, New York, 1972.
15. Ciarlet P. *The Finite-Element Method for Elliptic Problems.* Amsterdam, New York, Oxford: North-Holland Publishing Company, 1978.
16. Rektorys K. *Variational Methods in Mathematics, Science and Engineering.* Dordrecht, Holland/ Boston, USA/ London, England: Dr. Reidel Publishing Company; Prague, Chechoslovakia: SNTL Publisher of Technical Literature, 1980.
17. Forsythe G., Vasov W. *Finite-Difference Methods for Partial Differential Equations.* New York, London: John Wiley & Sons, Inc. 1959.
18. Godunov S.K., Ryaben'ki V.S. *Difference Schemes.* M.: Nauka, 1977 (in Russian).
19. Samarski A.A., Nikolaev E.S. *Methods for the Solution of Grid .* Equations. M.: Nauka, 1978 (in Russian).
20. Saul'ev V.K. *Integration of Parabolic Equations.* M.: Phyzmatgiz, 1960 (in Russian).
21. Fedorenko R.P. *Introduction to Computational Physics.* M.: Izdat. MPhTI, 1994 (in Russian).

Functional Analysis:

22. Vulikh B.Z. *Introduction to Functional Analysis.* M.: Gos. izdat. phyz. - mat. lit., 1958 (in Russian)
23. Lyusternik L.A., Sobolev V.I. *Short Course of Functional Analysis* M.: Vysshaya shkola, 1982 (in Russian).
24. Trenogin B.A. *Functional Analysis.* M.: Nauka, 1980 (in Russian).
25. Kantorovich L.V., Akilov V.A. *Functional Analysis.* M.: Nauka, 1977 (in Russian).
26. Yosida K. *Functional Analysis.* Berlin, Göttingen, Heidelberg: Springer–Verlag, 1965.
27. Trenogin V.A., Pisarevski B.M., Soboleva T.S. *Problems and Exercises in Functional Analysis.* - M.: Nauka, 1984 (in Russian).
28. Collatz L. *Functional Analysis und numerishe Matematik.* Berlin, Göttingen, Heidelberg: Springer – Verlag, 1964.
29. Varga R. *Functional Analysis and Approximations Theory in Numerical Analysis.* Philadelphia: SIAM, 1971.
30. Bakhvalov N.S., Zhidkov N.P., Kobel'kov G.M. *Numerical Methods* M.: Nauka, 1987 (in Russian).
31. Berezin I.S.m Zhidkov N.P. *Computational Methods.* M.: Gos. izdat. phyz.- mat. lit., 1959 (in Russian).
32. Babenko K.I. *Bases of Numerical Analysis.* M.: Nauka, 1986 (in Russian).
33. Volkov E.A. *Numerical Methods.* M.: Nauka, 1982 (in Russian).
34. D'yakonov E.G. *Minimization of Computations.* M.: Nauka, 1989 (in Russian).

35. Lanczos C. *Applied Analysis.* Englewood Cliffs, NJ: Prentice Hall, Inc., 1956.
36. Marchuk G.I. *Methods of Numerical Mathematics.* New York: Springer–Verlag, 2nd ed., 1982.
37. Mysovskikh I.P. *Interpolation Cubature Formulas.* M.: Nauka, 1981 (in Russian).
38. Sobolev S.L. Introduction to the Theory of Cubature formulas. M.: Nauka, 1974 (in Russian).
39. Hairer E., Wanner G. *Solving Ordinary Differential Equations.* II. Springer–Verlag, 1991.
40. Hageman L., Young D. *Applied Iterative Methods.* Academic Press, 1981.
41. Wachspress E. *Iterative Solution of Elliptic Systems and Applications to the Neutron Diffusion Equations of Reactor Physics.* Englewood Cliffs, NY: Prentice Hall, 1966.

The Author's Works Used in This Book:

42. Lebedev V.I. *The Composition Method.* M.: Dept. Num. Math. USSR Academy of Sciences, 1986.
43. Lebedev V.I. A new method for determining the roots polynomials of of least deviation on a segment with weight and subject to additional conditions. Part I, Part II. *Russ. J. of Numer. Anal. and Mathem. Modelling.* V. 8, N 3, 195–222; v. 8, N 5, 397–426, 1993.
44. Lebedev V.I. Chebyshev and optimal-in-mean iterative methods for problems with spectra contained in two segments or inside lemniscates. *East-West. J. Numer. Math.* V. 2, N. 2, 107–127, 1994.
45. Lebedev V.I. How to Solve Stiff Systems of Differential Equations by Explicit Method. In: Numerical Methods and Applications. Boca Raton, Ann Arbor, London, Tokyo: CRC Press, 45–80, 1994.
46. Lebedev V.I. Extremal polynomials with restrictions and optimal algorithms. *Advanced Mathematics: Computations and Applications.* Novosibirsk: NCC Publisher, 1995, 491–502.